D0368514

Groundbreaking Scientific Experiments, Inventions and Discoveries of the 17th Century

GROUNDBREAKING SCIENTIFIC EXPERIMENTS, INVENTIONS AND DISCOVERIES OF THE 17TH CENTURY ⎯⎯⎯⎯

MICHAEL WINDELSPECHT

Groundbreaking Scientific Experiments, Inventions and Discoveries through the Ages
Robert E. Krebs, Series Adviser

GREENWOOD PRESS
Westport, Connecticut • London

Library of Congress Cataloging-in-Publication Data

Windelspecht, Michael, 1963–
    Groundbreaking scientific experiments, inventions and discoveries of the 17th century
    / Michael Windelspecht.
        p.   cm.—(Groundbreaking scientific experiments, inventions and discoveries
    through the ages)
    Includes bibliographical references and index.
    ISBN 0–313–31501–9 (alk. paper)
    1. Science—History—17th century   2. Technology—History—17th
    century.   I. Title.   II. Series.
    Q125.W7917   2002
    509.032—dc21   2001023316

British Library Cataloguing in Publication Data is available.

Library of Congress Catalog Card Number: 2001023316
ISBN: 0–313–31501–9

First published in 2002

Greenwood Press, 88 Post Road West, Westport, CT 06881
An imprint of Greenwood Publishing Group, Inc.
www.greenwood.com

Printed in the United States of America

The paper used in this book complies with the
Permanent Paper Standard issued by the National
Information Standards Organization (Z39.48–1984).

10 9 8 7 6 5 4 3 2 1

For my mother—
may our memories of your love never fade

CONTENTS

LIST OF ENTRIES

SERIES FOREWORD

The material contained in five volumes in this series of historical groundbreaking experiments, discoveries, and inventions encompasses many centuries from the pre-historic period up to the twentieth century. Topics are explored from the time of pre-historic humans, the age of classical Greek and Roman science, the Christian era, the Middle Ages, the Renaissance period from the years 1350 to 1600, the beginnings of modern science of the 17th century, and great inventions, discoveries, and experiments of the 18th and 19th centuries. This historical approach to science by Greenwood Press is intended to provide students with the materials needed to examine science as a specialized discipline. The authors present the topics for each historical period alphabetically and include information about the women and men responsible for specific experiments, discoveries and inventions.

All volumes concentrate on the physical and life sciences and follow the same historical format that describes the scientific developments of that period. In addition to the science of each historical period, the authors explore the implications of how historical groundbreaking experiments, discoveries, and inventions influenced the thoughts and theories of future scientists, and how these developments affected peoples' lives.

As readers progress through the volumes, it will become obvious that the nature of science is cumulative. In other words, scientists of one historical period draw upon and add to the ideas and theories of earlier periods. This is evident in contrast to the recent irrationalist philosophy of the history and sociology of science that views science, not as a unique, self-correcting human empirical inductive activity, but as just another social or cultural activity where scientific knowledge is conjectural, scientific laws are contrived, scientific theories are all false, scientific facts are fickle, and scientific truths are relative. These volumes belie postmodern deconstructionist assertions that no scientific idea has greater validity than any other idea, and that all "truths" are a matter of opinion.

For example, in 1992 the plurality opinion by three jurists of the U.S. Supreme Court in *Planned Parenthood v. Case* restated the "right" to abortion by stating: "*at the heart of liberty is the right to define one's own concept of existence, of meaning of the universe, and of the mystery of human life.*" This is a remarkable deconstructionist statement, not because it supports the right to abortion, but because the Court supports the relativistic premise that anyone's concept of the universe is whatever that person wants it to be, and not what the universe actually is based on: what science has determined by experimentation, the use of statistical probabilities, and empirical inductive logic.

When scientists develop factual knowledge as to the nature of nature they understand that "rational assurance is not the same thing as perfect certainty." By applying statistical probability to new factual data this knowledge provides the basis for building scientific hypotheses, theories, and laws over time. Thus, scientific knowledge becomes self-correcting as well as cumulative.

In addition, this series refutes the claim that each historical theory is based on a false paradigm (a methodological framework) that is discarded and later is just superseded by a new more recent theory also based on a false paradigm. Scientific knowledge is of a sequential nature that revises, adds to, and builds upon old ideas and theories as new theories are developed based on new knowledge.

Astronomy is a prime example of how science progressed over the centuries. Lives of people who lived in the pre-historical period were geared to the movement of the sun, moon, and stars. Cultures in all countries developed many rituals based on observations of how nature affected the flow of life, including the female menstrual cycle, their migrations to follow food supplies, or adaptations to survive harsh winters. Later, after the discovery of agriculture at about 8000 or 9000 B.C.E., people learned to relate climate and weather, the phases of the moon, and the periodicity of the sun's apparent motion to the Earth as these astronomical phenomena seemed to determine the fate of their crops.

The invention of bronze by alloying first arsenic and later tin with copper occurred in about 3000 B.C.E. Much later, after discovering how to use the iron found in celestial meteorites and still later, in

1000 B.C.E. when people learned how to smelt iron from its ore, civilization entered the Iron Age. The people in the Tigris-Euphrates region invented the first calendar based on the phases of the moon and seasons in about 2800 B.C.E. During the ancient and classical Greek and Roman periods (about 700 B.C.E. to A.D. 100) mythical gods were devised to explain what was viewed in the heavens or to justify their behavior. Myths based on astronomy, such as the sun and planet gods as well as Gaia the Earth mother, were part of their religions affecting their way of life. This period was the beginning of the philosophical thoughts of Aristotle and others concerning astronomy and nature in general that predated modern science. In about 235 B.C.E. the Greeks first proposed a heliocentric relationship of the sun and planets. Ancient people in Asia, Egypt, and India invented fantastic structures to assist the unaided eye in viewing the positions and motions of the moon, stars, and sun. These instruments were the forerunners of the invention of modern telescopes and other devices that made modern astronomical discoveries possible. Ancient astrology was based on the belief that the positions of bodies in the heavens controlled one's life. Astrology is still confused with the science of astronomy, and it still is not based on any reliable astronomical data.

The ancients knew that a dewdrop of water on a leaf seemed to magnify the leaf's surface. This led to invention of a glass bead that could be used as a magnifying glass. In 1590 Zacharias Janssen, a spectacle-maker, discovered that two convex lenses, one at each end of a tube, increased the magnification. In 1608 Hans Lippershey's assistant turned the instrument around and discovered that distant objects appeared closer, thus the telescope was discovered. The telescope has been used for both navigation and astronomical observations from the 17th century up to the present time. The inventions of new instruments, such as the microscope and telescope, led to new discoveries such as the cell by Robert Hooke and the four moons of Jupiter by Galileo, who made this important astronomical discovery that revolutionized astronomy with a telescope of his own design and construction. These inventions and discoveries enabled the expansion of astronomy from an ancient "eyeball" science to an ever-expanding series of experiments and discoveries leading to many new theories about the universe. Others invented and improved astronomical instruments, such as the reflecting telescope combined with photography, the spectroscope, and Earth-orbiting astronomical instruments resulting in the discovery of new planets, galaxies, and new theories related to astronomy and the universe in the 20th century. The age of "enlightenment" through the 18th and 19th centuries culminated in an explosion of new knowledge of the universe that continued through the 20th and into the 21st centuries. Scientific laws, theories, and facts we now know about astronomy and the universe are grounded in the experiments, discoveries, and inventions of the past centuries, just as they are in all areas of science.

The books in the series Groundbreaking Experiments, Discoveries, and Inventions are written in easy to understand language with a minimum of scientific jargon. They are appropriate references for middle and senior high school audiences, as well as for the college level non-science major, and for the general public interested in the development and progression of science over the ages.

Robert E. Krebs
University of Illinois at Chicago

ACKNOWLEDGMENTS

I would like to recognize the invaluable assistance of several individuals who made this volume possible. First, I would like to thank Dr. Robert Krebs, who, as the editor for this series, provided invaluable assistance and guidance in the preparation of this volume. Next, I would like to thank Rae Dejur and Sandra Windelspecht for the design of the graphics used in this volume. Both of these individuals not only endured constant revisions, but also provided useful discussions on the artwork found within this work. Additional artwork was made available from the collections at the Library of Congress, National Library of Medicine, the National Space Science Data Center, and NASA. I wish to also thank my wife, Sandra, and my family and friends for their assistance and support.

Finally, I would like to thank Greenwood Publishing for recognizing the importance of this series. Science is not a modern creation; it is a culmination of centuries of dedicated work by countless scientists and mathematicians. To truly understand where science and technology is heading in the future, it is necessary not only to educate people on the rich history of science, but also to understand the thoughts of those who provided the foundations of our modern world.

INTRODUCTION

As is the case with any historical work, it is frequently difficult to establish a boundary of dates that conclusively defines the subject matter being studied. The history of science is no exception. Scientific discovery is a building process, with each new discovery not only founded on the work of previous generations of scientists, but also providing the material for later investigations and ideas. Such is the case when studying the scientific achievements of the 17th century. Science in the 1600s is the result of an effort that began prior to the Renaissance. Following an almost eleven-century lapse of activity, the Renaissance represents a time in the Western world of renewed interest in science and mathematics. This in no way implies that scientific investigation was in hibernation throughout the globe. In the centuries preceding the 17th century, Arabic scholars were studying the principles of optics and mathematics while Chinese astronomers were charting the paths of comets. However, scientific achievement is a periodic phenomenon, with civilizations demonstrating periods of high scientific achievement followed by times of relative inactivity. The 17th century represents an era of intense activity for Western scientists and while many nationalities contributed to the advance of science and math during this time, the focal points of activity developed primarily in the northern European countries and Italy.

The 17th century is a unique period of time in that it is possible to define the boundaries of scientific achievement during this century by examining the work of two scientists. In the early years of the 17th century (ca. 1600) the English physicist William Gilbert conducted preliminary investigations on the science of magnetism. While his work was important in characterizing magnetism as a force of nature, it represents a more important advance in that it was one of the first scientific studies to actively employ the use of the scientific method and experimentation to examine a force of nature. Gilbert's reliance on logic and experiments, rather than the philosophical basis of ancient Greek science, strongly influenced the science of Galileo, Johannes Kepler, and Isaac Newton. Over the next several decades, scientists in a wide variety of disciplines explored the nature of the physical world. By the end of the century (ca. 1687), their work enabled the English physicist and mathematician Isaac Newton to develop the theory of universal gravitation and the laws of motion. To accomplish this, Newton not only synthesized the work of his predecessors, but also co-developed a new form of mathematics (the calculus) to address the needs of his research. Newton himself stated: "If I have seen a little further it is by standing on the shoulders of Giants." In the history of science few centuries have produced such a wealth of scientific intellect as the 17th century. Together, within a period of less than 100 years, their accomplishments would change dramatically how scientists perceived the world around them.

The science of the ancient Greeks was based almost exclusively on observation of the natural world. In comparison, the scientists of the 17th century relied on experimentation and mathematical proofs to confirm their developing views of the natural world. As the scientists of this time advanced their studies, there developed a need for more advanced scientific instruments that would address specific research needs. To explore the heavens astronomers developed the telescope and, in doing so, first viewed the complexity of the heavens. With one generation of scientists examining the very large celestial objects, another used microscopes to discover the vast world of the microscopic organisms. As the century progressed more elaborate data-collecting devices were needed. The result was instruments such as the pendulum clock for keeping time and the barometer to measure air pressure. The link between science and technological advancement was thus established, a partnership that continues over four centuries later in modern science. However, unlike the 17th century, where advances in technology primarily benefited a select few of society, the technology of the 21st century has had a more significant impact, with few populations or areas of the globe not influenced by some technological or scientific achievement. Yet, the 17th century represents the time when the pace of technological advances in response to the needs of science was accelerated tremendously. The structure of our modern scientific culture is a direct descendant of these times.

Another important component of 17th-century science was the founding of scientific societies. These small groups of scientists, such as the Royal Society of London, provided a forum for the discussion of scientific topics. These often took the form of heated debates. Despite the considerable amount of controversy that these groups often generated, they remained a place to present new ideas for review by scientific peers. Furthermore, these societies frequently served as a clearinghouse for scientific information and provided an avenue for the publication of scientific findings.

One of the primary accomplishments of 17th-century science was the disproving of the theories and philosophies of the ancient Greeks. Although many scientists of the 1600s did not initially set out to refute the prevailing Greek ideas, and instead sought only to explain inconsistencies in the Greek theories, frequently the findings of the 17th-century scientific community completely uprooted established doctrines. While the scientific method, a process by which science is conducted, is common practice in the modern world, during the 17th century it was in its infancy. The ancient Greeks relied upon observations, supported by philosophical discussions, to establish theories on the physical laws of the universe. This changed dramatically in the 17th century as scientists not only attempted to develop scientific proof of their ideas using experiments, but were also forced to defend their positions against the scrutiny of their peers. This peer review of scientific findings is an important component of the modern scientific method.

All of the major scientific disciplines that were in existence at the start of the 17th century felt the effects of this change in methodology. In the study of astronomy Galileo's discovery of the moons of Jupiter provided some of the final evidence against an Earth-centered solar system. In the biological sciences William Harvey established that the blood circulates in the body and that the heart serves as a pumping mechanism for the circulatory system. These effectively ended the Greek reign on human anatomy and physiology set forth by Galen centuries earlier. In the physical sciences new theories of gravitation and laws of falling bodies, planetary motion, and motion replaced centuries-old Greek ideas. New scientific disciplines were also founded in the 17th century. The science of chemistry began as a challenge to the Greek concept of alchemy, or the transmutation of metals. The study of weather, called meteorology, developed around the invention of the barometer to measure air pressure and the thermometer to measure temperature. In mathematics, entirely new branches of study, such as analytical geometry and calculus, were developed to aid scientists in their explorations and discoveries. By the end of the 17th century few Greek ideas remained, and those that did were under close scrutiny by the scientific community. With the close of the 17th century the process of scientific thought based exclusively on observation, a Greek tradition, was effectively ended. Science had entered the experimental age.

This work is designed as a reference volume for anyone interested in obtaining an overview of the advances in mathematics and science during the 17th century. The topics in this volume were chosen from a historical perspective due to their impact on the development of science and mathematics. In each case, the work of these investigators eventually led to a scientific, mathematical, or technological breakthrough that shaped the course of science, as well as in many cases society, in the following centuries. Each entry provides a brief history of the topic dating from the time of the ancient Greeks. This allows the reader to more fully understand the scientific climate of the times, and thus appreciate the methodology by which the scientists approached their discoveries. As many of the advances in the 17th century were built on the achievements of many disciplines, each entry is cross-referenced to other topics in the volume that may provide an additional insight on the science of the times. Within each entry are descriptions of the experiments or discoveries and how this achievement influenced the science and culture of the times. While in the perspective of 17th-century science and culture some of these topics may have been so revolutionary as not to have made a direct impact on the people of the times, all of the chosen topics made an important contribution to the development of modern science in later centuries.

This work is targeted at the general science audience and as such an attempt has been made to describe as many of the scientific and mathematical discoveries as possible using common language. For those areas that require deeper understanding of technical and scientific terminology a glossary is provided at the end of the work with each entry highlighted in boldface when it is first used in the text. Also included for each entry is a bibliography of reference materials that will direct the reader to sources of additional information, with a complete list of references available at the end of the work. Indexes of subjects and names used within the work allow readers to quickly access desired information from this century. These reference materials make this volume attractive for secondary-school libraries, undergraduate higher-education colleges, and universities where students may be seeking general information on a specific experiment or discovery. In addition, community libraries that wish to possess a general reference volume on the 17th century, as well as anyone with an interest in science history, will find this work a useful addition to their collection.

TIMELINE OF IMPORTANT EVENTS

1590   Lippershey and Janssen invent the first optical system for observing objects at a distance. This optical system is the precursor of the telescope and microscope.

1600   Gilbert publishes *De Magnete*, which includes his thoughts on magnetism as a force of nature and his proposition that the Earth had properties similar to that of a magnet. This work also contains one of the first descriptions of the scientific method in the 17th century.

1603   Fabrici discovers valves in the veins of legs. This indicated that blood flowed one way in the circulatory system, an important contribution to Harvey's later models of blood circulation.

1604   Kepler's early studies of light indicate that it decreases in intensity with distance according to the inverse-square law.

1607   Halley's comet first appears in the 17th century.

1608   Lippershey invents the first primitive refracting telescope.

1609   Kepler proposes his first two laws of planetary motion. The first states that the orbits of the planets are elliptical and the second describes the speed of an object along this elliptical path.

1609   Although some nebula may be seen with the naked eye, the first examination of a nebula using a telescope occurs in this year by Galileo. Many other nebula are quickly detected and a debate begins on the nature of these astronomical objects.

1610   Galileo discovers the first four moons of Jupiter. These would later be named Io, Europa, Ganymede, and Callisto.

1610   Galileo invents a primitive temperature-measuring device, the precursor to the thermometer.

1612 Marius discovers the Andromeda nebula.

1614 Napier describes logarithms. Until the invention of the calculator in the 20th century, these remain the prime mechanism of conducting arithmetic operations with large numbers.

1614 Sanctorius begins some of the earliest scientific experiments on human metabolism and physiology.

1617 Napier invents Napier's Bones, an early calculating device.

1620 Although the scientific method had already been in use since the beginning of the century, the reasoning of the method was not formalized until Bacon's publication of *Novum Organum*, which included a comparison of inductive and deductive reasoning.

1620 Gunter invents the Gunter scale.

1621 The mathematical behavior of a refracted beam of light is described by Snell. This would later become known as Snell's Law.

1622 Aselli makes the first written descriptions of the lymphatic system while studying the canine digestive system.

1622 Oughtred invents the first slide rule.

1623 Bauhin publishes a botanical reference guide describing over 6000 plant species.

1623 Schickard invents the first calculating machine. This early machine performed addition operations only.

1624 Helmont first uses the term *gas* to describe a state of matter. He begins several experiments to describe the nature of gases.

1628 Harvey publishes *On the Movement of the Heart and Blood in Animals*, which outlines his theories on human blood circulation.

1629 Fermat develops analytical geometry, although this information is not published until 1679.

1632 Galileo makes a formal challenge to the belief of an Earth-centered (geocentric) universe with his publication of *Dialogue on the Two Chief World Systems—Ptolemaic and Copernican*.

1632 Galileo presents his Law of Falling Bodies.

1635 Cavalieri develops the method of indivisibles, a precalculus method of finding the area beneath a curve.

1635 Gellibrand demonstrates the magnetic poles of the Earth are not located directly at the north and south poles. The deviation from true north/south called magnetic declination. He also suggests that the location of the magnetic poles varies over time.

1637 Descartes publishes *Discourse on Method*. This contains his description of analytical geometry.

1637   Fermat describes what is now known as his Last Theorem, which states that for the equation $x^n + y^n = z^n$, there is no solution if $n > 2$ and x, y, z, and n are all positive integers. This remains unsolved until 1995.

1638   Gascoigne invents the first micrometer, although it remains relatively unknown until 1666.

1641   Ferdinando de' Medici invents the first liquid thermometer.

1642   Pascal invents the first calculating machine that performs both addition and subtraction operations.

1644   The Torricelli Tube is invented to initially study vacuums, but eventually gives rise to the invention of the first barometer.

1645   Guericke invents the first air, or vacuum, pump.

1647   Hevelius makes the first map of the moon.

1648   Pascal discovers that air pressure varies with altitude.

1651   Highmore publishes some of the first studies of embryology using a microscope.

1651   Harvey publishes *De Generatione Animalium* discussing his ideas on the generation of animals.

1651   Riccioli makes a map of the moon's surface. Many of the modern names of surface features are derived from this map.

1652   Bartholin suggests that the lymphatic system in humans is actually a second circulatory system within the body.

1653   Pascal invents the Pascal triangle (first published in 1665) as part of his work on probability and binomials.

1654   Pascal and Fermat collaborate on studying probability theory in mathematics.

1654   Kircher correctly suggests that the moon is responsible for ocean tides, although he incorrectly identifies the force as magnetism.

1656   Huygens discovers the rings of Saturn, first described by Galileo in 1612 as a bulge on the planetary disk.

1656   Huygens develops a model of the pendulum clock that has an increased level of accuracy. This finds wide-scale application in scientific studies that require more precise time measurements.

1658   Swammerdam discovers red blood cells.

1659   Huygens discovers the first surface features on Mars.

1660   Hooke discovers that air pressure varies with oncoming storms, a first step for the developing science of meteorology.

1660   Boyle experimentally demonstrates that respiration is required for life.

1660   Guericke conducts some early scientific experiments on static electricity.

1661   Malpighi first detects the capillaries of the circulatory system in the wing of a bat.

1661   Boyle publishes *The Skeptical Chemist*. This work made significant advances in the study of both chemistry and the physical laws of air pressure.

1662   Boyle's Law on the relationship between pressure and the volume of a gas is presented. This was important not only in the study of gases, but also in the developing theories of the nature of atoms and elements.

1662   Wren introduces the first rain gauge into Western meteorology.

1663   Somerset invents a water-commanding device. This instrument uses steam to create a vacuum for the purpose of moving water. This is considered to be the first functioning steam-powered machine.

1664   Hooke discovers the Great Red Spot of Jupiter.

1665   Hooke publishes *Micrographia* containing some of the first descriptions of microscopic cells.

1665   Grimaldi's studies on the diffraction of light waves are published.

1665   Cassini determines that Mars rotates on its axis, and he calculates the rotation of the planet.

1665   Newton begins his experiments on light, studying the nature of color. His studies indicate that color is a property of light, and not the object.

1666   Auzout invents the first widely distributed micrometer. This instrument has an important impact on the measuring of astronomical distances.

1668   Redi conducts a series of experiments that disprove the concept of spontaneous generation of life.

1668   Leeuwenhoek first notices the movement of red blood cells in capillaries, thus completing the model of the circulatory system begun by Harvey (see 1628).

1668   Hevelius publishes *Cometographia*, which contains some of the most detailed descriptions of comets to date.

1668   Newton builds the first reflecting telescope.

1669   Becher develops the phlogiston theory of combustion.

1669   Brand discovers phosphorus.

1669	Steno presents the idea that fossils are the petrified remains of organisms that lived on the planet in the distant past.
1672	Cassini, Picard, and Richer determine the distance to Mars using a process called parallax analysis. This was one of the first indications of the true dimensions of the solar system.
1672	Newton presents his theories on color and light.
1674	Leibniz invents the first calculating machine that performs multiplication and division operations.
1674	Perrault presents one of the first modern discussions on the origin of springs, a precursor to the science of hydrology.
1675	Leeuwenhoek discovers protozoans, which he calls *animacules*, in pond water.
1676	Roemer makes one of the first calculations on the speed of light. He estimates the speed of light to be around 140,000 miles per second.
1677	Leeuwenhoek first identifies sperm cells.
1678	Hooke releases his discoveries on the theory of elasticity, commonly called Hooke's Law.
1678	Halley constructs one of the first star atlases of the Southern Hemisphere.
1679	Leibniz develops the binary number system.
1679	Papin invents the *digester*, a form of pressure cooker.
1680	Borelli contributes to the knowledge of human physiology when he suggests that the muscles and bones function in the same manner as levers.
1682	Halley's comet appears for the second time in the 17th century.
1682	Grew demonstrates that plants reproduce by sexual reproduction.
1683	Leeuwenhoek discovers bacteria.
1684	Leibniz independently develops the calculus as a mathematical tool.
1684	Newton publishes *De Motu*. This work is the foundation for Newton's later studies of gravitation and motion.
1686	Ray begins publication of *Historia Plantarum*. This work not only includes taxonomic descriptions of plants, but also is one of the first attempts at defining a species.
1687	Newton publishes his description of the calculus.

1687   Newton publishes *Principia*, which contains revolutionary discussions on the nature of light, the theory of universal gravitation, the shape of the Earth, the laws of motion, and the nature of ocean tides. This one work summarizes much of the scientific achievement of the 17th century.

1690   Huygens publishes his ideas that light exists as a wave. This opposes Newton's philosophy that light exists as a particle.

1691   Bernoulli invents a polar coordinate system for use in analytical geometry.

1693   Leibniz invents a calculating machine that performs multiplication and division operations.

1699   Amontons invents the air thermometer.

1704   Newton publishes *Optiks*, which contains a complete account of his theories of the physical properties of light.

EXPERIMENTS, INVENTIONS AND DISCOVERIES

A

Analytical Geometry (ca. 1629–1637): Geometry may be defined as the branch of mathematics that studies the two- and three-dimensional positioning of shapes (circles, triangles, etc.). Derived from the Greek words for *earth* and *measurement*, geometry historically was used in the surveying of land. Architects throughout time have also used geometrical thinking in the design and construction of buildings and temples. While these uses of geometry predate the rise of the Greek civilization, it was the ancient Greeks who were the first to apply the processes of logical thinking to the study of geometry. One of the earliest, and most recognized, examples of Greek geometry was provided by Pythagoras (ca. 580 B.C.E.). Pythagoras formally defined the relationship between the hypotenuse and sides of a right triangle, although the Babylonians had understood this relationship centuries earlier. However, the major advances in Greek geometry may be attributed to the work of Euclid (ca. 330–ca. 270 B.C.E.). Euclid pioneered the use of an axiomatic system of geometry. In this system a series of true statements are established, called **axioms**, which are in turn then used to develop general geometric **theorems**. Euclid's *Elements* (ca. 300 B.C.E.) was a compilation of Greek mathematics to date and listed hundreds of such theorems. *Elements* formed the basis of geometric thinking until the time of the Renaissance.

To the ancient Greeks geometry was a tool to examine algebraic relationships, a process sometimes called geometric algebra. However, in order to completely understand the properties of a geometric shape, it is first necessary to be able to define and manipulate the geometric figure in terms of an algebraic equation. This form of mathematics is called **analytical geometry**. The first step in identifying the algebraic

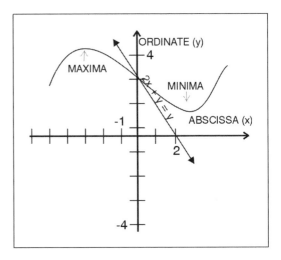

Figure 1. The use of a coordinate system to plot the solution to the equation $2x + y = 4$. This was an important step in the development of analytical geometry in that it allowed the solution, as well as the x- and y-axis intercepts, to be visualized. The curve represents Fermat's principles of minima and maxima.

parameters of a shape is to define its location, or coordinates, in space. The use of coordinate systems was nothing new to mathematics. The ancient Greek mathematicians Archimedes (ca. 287–212 B.C.E.) and Apollonius of Perga (ca. 262–ca. 190 B.C.E.) defined the location of an object using a coordinate system derived from the object's **longitude, latitude**, and height above the surface. Later the Greek astronomers Hipparchus (ca. 146–ca. 127 B.C.E.) and Ptolemy (ca. 100–ca. 178) utilized similar strategies when constructing early stellar maps (see STAR ATLASES). The application of coordinates to algebraic functions began in the 14th century with the work of the Frenchman Nicholas Oresme (ca. 1323–1382), who first used coordinates to graph dependent variables against independent variables. However, it was not until the 17th century that the French scholars René Descartes (1596–1650) and Pierre de Fermat (1601–1665) would independently develop a coordinate system that could be applied to the developing field of analytical geometry. In this system the coordinates are established by the relationship of the point with regard to two perpendicular axes, the **abscissa** (the x-axis) and the **ordinate** (y-axis). For example, in the equation $2x + y = 4$, if x is equal to 3, then y equals -2. Each solution for a specific value of x yields a corresponding value of y. If these points are plotted using a coordinate system, it is possible to visualize the geometric solution of $2x + y = 4$ (see Figure 1). While Fermat was most likely the first to develop such a method, it was Descartes who published his ideas first. For this reason our current coordinate system is now commonly called a Cartesian system. However, the modern terms abscissa, coordinate, and ordinate are not of Cartesian origin;

instead they were first used around 1692 by the German mathematician Gottfried Leibniz (1646–1716) in his development of calculus. Once the coordinate system established the location of an object using coordinates it became possible to determine the formula of a line or curve passing through a point on the geometric shape. This formed the basis for much of the work on analytical geometry in the 17th century.

In 1637 Descartes published *Discours de la Méthode* (*Discourse on Method*). Contained within this work were the author's viewpoints on the philosophy of science and the methods of scientific reasoning (see SCIENTIFIC REASONING). Attached to it were three appendices designed as case studies to support the philosophy presented in *Discours.* Two of the volumes discussed optics and meteorology (see METEOROLOGY; OPTICS), while the other volume, *La Geometrie,* was dedicated to the presentation of Descartes' analytical geometry. The material contained within *La Geometrie* made several significant improvements to mathematical thinking. Perhaps the most important among these was the development of mathematical symbolism in algebraic equations. Descartes was one of the first to use letters at the beginning of the alphabet to represent known numerical quantities (a,b) and letters from the end of the alphabet to designate unknown quantities (x,y). An excellent example of this type of symbolism is the equation for a slope of a line:

$$y = ax + b$$

In this equation the variables a and b represent unknowns while the y and x represent coordinate locations. Furthermore, Descartes was a pioneer in the development of exponential notation (x^2, x^3), a problem that had consistently plagued Greek mathematicians. While these advances may not seem significant in modern thinking, the establishment of these rules was an important precursor in the development of analytical geometry. Descartes was convinced that the combination of algebraic and geometric mathematics would provide a powerful tool in establishing standardized equations defining the slope of lines, **tangents** to a circle, and so on. In fact, our modern analytical geometry uses algebra to solve geometric problems, and vice versa, an ability first effectively employed by Descartes.

La Geometrie is divided into three major sections, one of which covers coordinate systems. A second section uses coordinate geometry to classify curves, including a method of constructing tangents to curves. Tangents, straight lines that intersect a curve or circle at only a single point, were known prior to the 17th century. However, the coordinate system allows the exact location of that point to be plotted. Thus, Descartes could use tangents to define the algebraic, rather than geometric, properties of the curve. The third section explores the solution of equations of a degree higher than two (for example, x^3, x^4). Greek geometry had been limited to solving equations with degrees less than two. To facilitate the solution of these high-level polynomials, Descartes developed a series of algebraic standards, called the "Rule of Signs." The Rule of Signs allowed for the solving of complex roots of equations by establishing the maximum number of positive and negative roots in a **polynomial**. Included in the Rule of Signs is Descartes' standardization of symbolism noted earlier.

While historically regarded as a revolutionary work, *La Geometrie* was difficult to decipher and frequently did not illustrate the methods by which Descartes solved problems. A more important influence on 17th-century mathematics came from Pierre de Fermat. Fermat appears to have developed analytical geometry as early as 1629, although publication of his work *Introduction to Plane and Solid Loci* did not occur until 1679, several years after his death. While often considered coinventors of analytical geometry, in reality Fermat's solutions and political views differed significantly from those of Descartes. Fermat and Descartes openly disagreed on the methods of determining tangents to curves, as well as the means for calculating the minima and maxima of equations. As their names imply, the minima and maxima represent the extreme solutions for an equation (see Figure 1). In the posthumous publication *Method of Finding Minima and Maxima*, Fermat standardized the means of algebraically determining these values for geometric figures. This method closely parallels the process of **differentiation** used in modern calculus.

In addition, Fermat's algebraic work on the equations of curves allowed him to further define the equations for a number of geometric shapes. Included in *Introduction to Plane and Solid Loci* were the solutions for a hyperbola ($a^2 = xy$), parabolas ($ay = x^2$), and a circle ($x^2 + y^2 + 2ax + 2by = c^2$). He was interested, as were many other mathematicians of the 17th century, in algebraically finding the area under a curve.

Fermat's analytical geometry most closely resembles modern methods and served as the basis for much of the work later in the 17th century on calculus (see CALCULUS). However, his work was not limited to geometry. In collaboration with Blaise Pascal (1623–1662) he made significant contributions to studies on probability (see PROBABILITY), a precursor of modern statistics. Fermat also contributed to developing ideas on number theory. A portion of his work, Fermat's Last Theorem, remained unsolved until the end of the 20th century (see FERMAT'S LAST THEOREM).

Together with the invention of logarithms (see LOGARITHMS), the advances in analytical geometry early in the 17th century marked a reinvention of mathematical thinking. Fermat and Descartes were not the only contributors to this field during this century. For example, in 1691 Jakob Bernoulli (1654–1705) developed a polar coordinate system that uses as one axis the degree of an angle, with distance represented on the second axis. Others such as Gregory of St. Vincent (1584–1667) and Evangelista Torricelli (1608–1647) further investigated using analytical geometry to determine the area under a curve. In fact, by the close of the 17th century the majority of Western mathematicians were involved in some aspect of using algebra for the solving of geometric problems. More importantly, analytical geometry formed the basis for the development of calculus by Leibniz and Isaac Newton (1642–1727) later in the 17th century. This progression of mathematical thought correlates directly with other scientific advances of this time. Without the evolution of mathematical methods early in the century, later unifying theories in areas such as gravity would not have been possible. The fact that analytical geometry was in work simultaneously by two independent researchers is a further indication that the scientific climate of the early 1600s was ripe for such a discovery.

Selected Bibliography

Boyer, Carl B., and Uta C. Merzbach. *A History of Mathematics.* 2nd ed. New York: John Wiley & Sons, 1989.

Cooke, Robert. *The History of Mathematics: A Brief Course.* New York: John Wiley & Sons, 1997.

Katz, Victor J. *A History of Mathematics: An Introduction.* Reading, MA: Addison-Wesley Longman, 1998.

Kline, Morris. *Mathematical Thought from Ancient to Modern Times.* New York: Oxford University Press, 1972.

Stillwell, John. *Mathematics and Its History.* New York: Springer-Verlag, 1989.

Animal Generation (1668–1677): Throughout recorded history humans have always held an interest in the process of procreation. While in the modern world the role of the male and female sex cells, or **gametes**, in the process of reproduction is clearly understood, the lack of accurate scientific instruments in ancient civilizations resulted in the formation of many creative theories on how animals were generated. The longest lived of these theories, dating from before the time of the ancient Greeks, was the concept of spontaneous generation. In the process of spontaneous generation, living creatures are derived from inorganic, or nonliving, materials.

This idea was widely accepted until the mid-1600s when it was finally challenged by experimental procedures. In 1668, the Italian Francesco Redi (1628–1694) designed a series of experiments to test the validity of spontaneous generation. Prior to Redi's experiments it was widely believed that maggots were generated from rotting meats. However, Redi was convinced that flies visiting the decaying material were contributing to the formation of the maggots. To test his hypothesis, Redi prepared a series of jars that contained various types of meat. Half of the jars were sealed while the other half remained exposed to the environment. After a period of time, only the jars that were left exposed revealed the presence of maggots. Based on these results, Redi concluded that the maggots were not developing directly from the decaying meat as was previously suggested by believers of spontaneous generation. However, the possibility still existed that the maggots were being generated from fresh air and subsequently being deposited on the meat. This would offer a possible explanation as to why the meats in the sealed jars did not generate maggots. To test the contribution of this second variable, Redi once again placed meats in two sets of jars. As with the previous experiment, he left one set exposed to the environment; however, this time the second set was sealed with gauze. The gauze allowed fresh air to reach the meat, but not the flies. Once again his results indicated the presence of maggots on the exposed meat, but not on the meat contained within the jars covered with gauze.

Redi's experiments were significant during the 17th century for two reasons. First, his results had indicated that the flies directly contributed to the formation of the maggots on the meat, and not the decaying material or inorganic air. While Redi suggested that the maggots were formed from eggs deposited by the flies, it would not be until two decades later that the Dutch biologist Anton van Leeuwenhoek (1632–1723), using one of the new generation of powerful microscopes (see MICROSCOPE), confirmed Redi's experiments. However, from a larger perspective, Redi's experiments were important in that they are one of the first documented instances of the use of controls in the biological sciences. Controls are an important part of the experimental method in that they allow a scientist to verify that the experiment was designed and executed successfully. By leaving a set of jars exposed to the environment, Redi was able to confirm that maggots would form in the time frame of his experiment. Today, the design of experimental procedures with precise controls is an integral part of scientific investigations.

Redi was not the only investigator to challenge the belief in spontaneous generation. In early investigations of plant-insect interactions, Antonio Vallisnieri (1661–1730) had described that certain species of insects did not generate spontaneously from plant material, but rather were formed from eggs deposited by adult members of the species. While both the Redi and Vallisnieri experiments challenged the validity of spontaneous generation, and suggested the role of the egg in reproduction, they did not venture to explain the formation of the egg as a reproductive mechanism. The improvement in the magnification power of microscopes in the 17th century eventually led to the discovery of cells and microorganisms (see CELLS). This resulted in new ideas and hypotheses on the process of animal generation. And as the 17th century drew to a close, two main camps were forming in the debate, those of the preformists and the epigenesists.

The predominant theory of the time was that of **preformation**, or the idea that the embryo was completely formed, with all structures existing as miniature versions of their final form. The study of **embryology** was not new to the 17th century. The Greek philosopher Aristotle (384–322 B.C.E.) published information on the embryonic development of chickens centuries earlier (see EMBRYOLOGY). For the preformists, the embryonic development simply consisted of a gradual enlarging of the size of the structures. To a preformist, each embryo contained the preformed embryo of the next generation inside of it, which contained the next generation inside of it, and so on to infinite levels. Noted scientists such as Marcello Malpighi (1628–1694) and Anton van Leeuwenhoek actually contributed to these theories by claiming to have observed miniature preformed creatures during their observations with microscopes (see EMBRYOLOGY). While the preformists acknowledged that the males and females both make some contribution to the development of the embryo, they differed in where the preformed embryo resided. Eggs had been previously identified in many animals, and their role in reproduction was well established. However, it was not until 1677 when Leeuwenhoek first identified sperm cells that the possibility that the males housed the preformed embryo began to take shape. To the animalculists, the egg was simply provided for nourishment of the embryo inside of the sperm cell, while the ovists believed that the sperm cells simply activated the development of the embryo inside of the egg cell. The conflict between the ovists and the animalculists would last well into the next century when more experimental, and less philosophical, investigations would occur into the process of embryology.

A second theory that took shape during the 17th century was the concept of **epigenesis**. In epigenesis, the structures of the embryo were gradually formed from a simpler substance. One of the first to publish experimental data in favor of epigenesis was William Harvey (1578–1657). Harvey, well known for his studies of the circulatory system (see BLOOD CIRCULATION), also spent time investigating the development of embryos within chicken eggs. By examining the physical characteristics of embryos within chicken eggs over a period of time, Harvey was able to describe the developmental progress of the embryo. Harvey observed that the developmental pathway of animals started as a single point of undifferentiated matter, and then gradually developed into a complex organism. Harvey's descriptions, published in 1651 in *De Generatione Animalium*, would do little to influence the position of the preformists of his time. However, it was later considered to be an important factor in the developing theories of embryology (see EMBRYOLOGY).

Thus, while theories on the process of animal generation at the start of the 17th century favored the concept of spontaneous generation, it was recognized by the end of the century that life arose from cells within the male and female of the species. In the 18th century a third theory, **pangenesis**, suggested that all the parts of the body made minor contributions to the seed, which in turn became a new embryo (see EMBRYOLOGY). Although there would not be a resolution on the role of the male and female gametes by the end of the century, even though the limited available information favored the theory of preformation, these studies were an important break from the philosophical beliefs of the preceding thousand years. The difference in opinions between the epigenesists and the preformists actually benefited scientific discovery in later times as each group attempted to collect experimental evidence that supported its position. It was not until the 18th century that a consensus would be reached and the principles of the generation of animals resolved.

Selected Bibliography

Gasking, Elizabeth. *Investigations into Generation: 1651–1828*. Baltimore, MD: Johns Hopkins University Press, 1967.

Harvey, William. *Anatomical Exercises on the Generation of Animals.* In *Great Books of the Western World,* vol. 28, edited by Robert M. Hutchins. Chicago: Encyclopedia Britannica, 1952.

Pinto-Correia, Clara. *The Ovary of Eve: Egg and Sperm and Preformation.* Chicago: University of Chicago Press, 1997.

Serafini, Anthony. *The Epic History of Biology.* New York: Plenum Press, 1993.

Atomic Theory (ca. 1661): The debate on the existence of atomic particles had its origins long before the invention of present-day scientific methods. In modern science the term **atom** is defined as the smallest part of an element that retains the chemical properties of that element. However, the original definition, first developed by the Greek philosopher Democritus (ca. 460–ca. 370 B.C.E.), means *indivisible.* Democritus, and his mentor Leucippus (ca. 5th century B.C.E.), believed that matter consisted of infinite uniform particles that were indestructible. These particles traveled through the void of space and could not be compressed by ordinary means. Epicurus (ca. 341–ca. 270 B.C.E.) extended this philosophy by stating that the entire universe was exclusively composed of atoms and the voids between the atoms, although it was the existence of the voids themselves that created debate among philosophers such as Aristotle. The question as to whether **matter** consisted of discrete particles or was continuous would prove to be a long-running debate. Furthermore, like many Greek scientists, Leucippus and Epicurus confined support for their theories to philosophical arguments only. Democritus, on the other hand, was far ahead of his time when he predicted that the activity of the senses, the sense of smell in particular, was due to the physical arrangement of atomic particles. But without the backing of scientific experimentation, a process that would not be firmly established in scientific investigations until the work of the 17th-century English physician William Gilbert (1544–1603), the concept of an atomic world would not gain widespread support (see MAGNETISM).

While no one single event in the 17th century proved the existence of atoms, a number of scientists were accumulating evidence on the atomic nature of matter as a result of research in other fields. One of the earliest and strongest supporters of atomic theory in the 17th century was the French philosopher Pierre Gassendi (1592–1655). Gassendi was a firm supporter of Epicurus's concept of atoms and voids. Gassendi also believed that this was the force that held the physical universe together, and he had a strong influence on the thinking of other atomists of the 17th century. While Gassendi was not an experimenter, he did make several important contributions. First, while many of the Greeks had simply regarded atoms as theoretical mathematical points, Gassendi considered them to have fundamental physical characteristics such as weight, dimensions, and shape. Like the Greeks before him, Gassendi considered atoms to be indivisible and indestructible. Second, Gassendi proposed that the larger compounds, which he called molecules, were formed from the interactions between atoms. While this may seem a logical conclusion in our modern scientific world, the chemistry of the 17th century had not yet advanced to this stage and as such Gassendi's ideas would remain largely unrecognized until the 19th century.

As the seeds of scientific inquiry developed in the early years of the 17th century, many experimental scientists were becoming increasingly dissatisfied with the ancient Greek explanations of the physical world, specifically the nature of **elements** and atoms. Aristotle had classified all matter as consisting of combinations of four elements—earth, fire, water, and air. This four-element theory persisted intact for almost two millennia. A number of early works began as researchers started to define the physical nature of the elements. Primary among these was the research into the physical properties of air pressure. In the 17th century the English chemist Robert Boyle (1627–1691) made two major advances that challenged Aristotle's philosophy. The first was in the definition of an element. Boyle simplified the definition of an element to indicate any matter that may not be broken up into smaller units of matter. The atom once again became the smallest of these units. He then defined a compound as being an object that could be divided. But more importantly, Boyle demonstrated the existence of atoms and the surrounding voids through his experiments with air pressure. In his experiments Boyle trapped a bubble of air between a column of mercury and the sealed end of a J-shaped tube (see BOYLE'S LAW for figure). As the amount of mercury was increased, the space that the air occupied at the end of the tube decreased. This was only possible if the void spaces around the individual atoms of air were being compressed. Boyle had provided the experimental verification of Epicurus's theories. Others, such as Blaise Pascal, Evangelista Torricelli, and Edme Mariotte (ca. 1620–1684), provided additional confirmations through their work on air pressure (see BAROMETER; BOYLE'S LAW).

While Gassendi and Boyle were the primary factors in the atomist movement of the 17th century, there were other contributors; however, little of their work would be experimental. The Italian scientist Galileo Galilei's (1564–1642) strong beliefs in the atomic structure of Democritus and Epicurus, coupled with his views on the structure of the solar system (see JUPITER), placed him in direct conflict with the ruling Catholic Church and created significant problems for him later in life. The English physicist Isaac Newton was also a supporter of atomic structure, including both the atom and the void, although he held a much less confrontational stance with the Church regarding his theories. More importantly, although Newton spent a considerable amount of his time developing unifying theories of gravitation (see GRAVITATION), he recognized that his own theories would not hold true at the atomic level. He proposed that at this level of organization **electromagnetic** forces were most likely the unifying force. This has since been proven by modern scientific methods.

Not all scientists of the 17th century were atomists. The French philosopher René Descartes belonged to a group called the antiatomists. Many of the Greeks who supported atomic ideas, Democritus for example, often considered atoms to be merely mathematical representations of matter, and not physical entities. Descartes expanded this notion into a theory of a **vortex** in which all of the atoms of matter were basically in one continuous motion with one another. Descartes developed elaborate theories on vortexes and it is known that his philosophy did have an impact on the thinking of many important minds of the 17th century, including Newton, who spent a considerable amount of time disproving vortex theories.

The development of atomic theory was not limited to the Western world. The scientific community of India also believed in the indivisible and indestructible nature of the atom. Scientists from this region frequently classified atoms as belonging to one of four classes. Different effects were caused by making various combinations of these classes. In Western society, discussions of atomic structure did not significantly impact the science of the 17th century. Scientists of the time lacked instruments to make detailed observations and measurements and had yet to develop advances in chemistry to support their ideas. A unifying atomic theory would have to wait until 1803 when John Dalton (1766–1844) proposed the three major components of the atomic theory, that is,

1. All matter is composed of atoms.
2. Atoms of a given element possess unique characteristics and weight.

3. Three types of atoms exist—elements, molecules, and compounds.

While Dalton eventually expanded his theory to discuss ratios of atoms in compounds and the atomic weight of elements, the experimental roots of Dalton's work began in the 17th century. The philosophical work of Gassendi and Boyle's research on air pressure set the stage for atomic thought in the following centuries. By the 20th century, the scientific advances in physics, coupled with vastly improved instrumentation, gave physicists the ability to examine the structure of the atom in considerable detail. The fields of atomic chemistry, which addresses the power of the atom as it relates to electrons, and nuclear chemistry, involving the structure of the atomic nuclei, are direct descendants of this work. Several centuries later, this ushered in the current Atomic (sometimes called Nuclear) Age, a time in which mankind has the ability to harvest the power of the atom.

Selected Bibliography

Booth, Verne H. *The Structure of Atoms.* New York: Macmillan, 1964.

Lloyd, G.E.R. *Early Greek Science: Thales to Aristotle.* New York: W. W. Norton, 1970.

Pullman, Bernard. *The Atom in the History of Human Thought.* New York: Oxford University Press, 1998.

B

Barometer (1644–1698): During the 17th century there were three major inventions and discoveries that focused on the nature of air. These were the vacuum pump, Boyle's Law on air pressure, and the barometer (see BOYLE'S LAW; VACUUM PUMP). Together, these instruments allowed scientists to finally examine the properties of air and address some of the historical questions regarding the nature of a **vacuum**. To the ancient Greeks a vacuum was an unnatural phenomenon. In fact, the Greek philosopher Aristotle believed that *nature abhors a vacuum* and that the void spaces between objects were filled with an element called **ether**, a fine compound that could pass through virtually all other materials. While it was recognized by the ancient Greeks that air possessed weight, they did not understand the concept of pressure (see BOYLE'S LAW). Such misconceptions would not be confined to the Greeks, as the renowned scientist Galileo appears not to have understood air pressure as late as the 17th century.

In the early part of the 17th century scientists remained in conflict as to the very existence of a vacuum. Early experiments focused primarily on the creation of vacuums and not on direct measurement of air pressure. Several experiments, most notably the one performed by the Italian physicist-astronomer Gasparo Berti (ca. 1600–1643), utilized air pumps to create vacuums at the top of tubes (see VACUUM PUMP). The first steps toward the creation of a device to study air pressure, the barometer, would begin with experimentation by the Italian scientists Evangelista Torricelli and Vincenzo Viviani (1622–1703). As with many scientific discoveries, it was an industrial problem that promoted the development of the device. For some time workers in mines had been experiencing problems removing water from the bot-

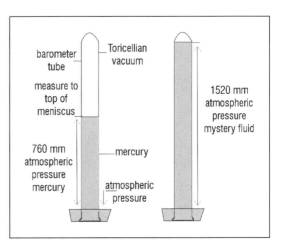

Figure 2. Torricelli's experiment to test the effects of air pressure on a column of mercury. This is frequently regarded as the prototype of the scientific instrument known as the barometer (courtesy of Greenwood Press).

toms of mine shafts. They had observed that the maximum height that they could raise a column of water by using a pump was about thirty feet. There were two major theories as to why this occurred. The first, supported by Galileo, was that the force of the vacuum within the tube regulated the maximum height of the water. Opposing this was the idea, proposed by the Dutch physicist Isaac Beeckman (1588–1637) and others, that it was the external pressure of the air that was the limiting factor in the maximum height of the water column.

Galileo is believed to have been one of the first to suggest an experiment that tested the relationship between the density of the liquid and the height of the column. The liquid of choice was liquid mercury (quicksilver). Mercury has a density 13.5 times that of water and it is one of the few metals that is a liquid at room temperature. Although Galileo is often credited with the idea, after Galileo's death it was Viviani who suggested the experimental design to a former student of Galileo, Evangelista Torricelli. In 1644 Torricelli designed a device to test the effects of air pressure on liquid mercury. The device was simple and consisted of a tube that was sealed at one end. The tube was filled with liquid mercury and then inverted into a container containing additional mercury (see Figure 2). Rather than draining from the sealed tube completely, a thirty-inch column of mercury remained suspended within the tube. Since the mercury is 13.5 times denser than water, it could be reasoned that the supported column would be 13.5 times smaller for the mercury. As the column of water was 33 feet (396 inches), and the resulting column of mercury was 30 inches, the Torricelli experiment supported the hypothesis. However, there remained some dispute as to what held the mercury in the tube. Viviani suggested that it was the weight of the atmosphere on the mercury in the open container that supported its weight, one of the first suggestions that the atmosphere had weight (see BOYLE'S LAW). Furthermore, as the mercury drained, a vacuum was formed between the mercury and the sealed end. While the "Torricelli Tube" was in fact a primitive barometer, it was primarily used to demonstrate the creation of a vacuum rather than to quantify changes in the pressure of air, although Torricelli proposed that the device would be useful for detecting variations in air pressure.

Blaise Pascal is credited with creating the first true barometer. Pascal fitted the Torricelli Tube with a marked scale to measure the changes in air pressure. With this instrument he was able to notice that the level of the liquid within the barometer changes with variations in the weather. Pascal used the barometer for many purposes, including his studies on the relationship between air pressure and altitude (see BOYLE'S LAW). Pascal had an intense interest in the properties of liquids and found the barometer to be an exceptionally useful instrument in the study of **hydrostatics** and **hydraulics**, fields in which he made significant contributions (see HYDROMECHANICS). Another investigator of hydrostatics was the Englishman Robert Hooke (1635–1702). Hooke was interested in using the barometer to further understand the structure and nature of the atmosphere. In addition, he made similar observations as Pascal regarding the changes of air pressure during storms and conducted some of the first weather-related measurements. This served as the foundation for the quantification of weather using specialized scientific instruments, the predecessor to modern meteorological science (see METEOROLOGY). It was Hooke who is believed to have first applied the term *barometer* to the instrument.

There were several improvements made to the barometer in the decades following the invention of the Torricelli Tube, many of these made by Robert Hooke. Many of these improvements focused on increasing the accuracy of the device so that it could be used for reliable quantitative scientific measurements. The mercury reservoir of the Torricelli Tube made the instrument difficult to handle and the instrument had to be operated on a perfectly level surface to function correctly. One variation, called a siphon barometer, angled the bottom of the tube upward to make the lower portion of the tube itself the reservoir. Hooke also developed a wheel barometer in which a cork was floated on the mercury. The small changes in the mercury level were amplified by attaching a string to the cork. The string was then run over a wheel (or pulley) and connected to a long pointer. Variations in the level of the barometer caused the tension of the string to change. The wheel and pointer amplified the changes, thereby enhancing the accuracy of the instrument. Hooke is also credited with inventing the highly accurate, at least for the 17th century, marine barometer. Edmund Halley (1656–1742) used one of these instruments in his charting of magnetic declination in the Atlantic Ocean (see MAGNETIC DECLINATION; MAGNETISM). Various experiments were also done by a number of investigators with multiple liquid barometers. Some, such as the German Otto von Guericke (1602–1682), replaced the mercury with water. Unfortunately, many of these devices were sensitive not only to air pressure, but also to temperature. This severely limited their accuracy. It would not be until the 18th century that more sensitive devices would be invented by the Swiss geologist Jean-André Deluc (1727–1817). These instruments would consist of purified mercury and leveling modifications to make them more transportable and less susceptible to temperature changes. Deluc developed a method of using the barometer to measure the heights of mountains.

Most modern barometers do not contain mercury, which being a heavy metal is toxic to humans. There were some interesting early attempts at designing mercury-free devices. Toward the end of the century both Pascal and Hooke experimented with the use of dead weights and bellows to detect changes in air pressure. However, these instruments were very large and unwieldy and thus had a limited level of accuracy. Another approach was the invention of the aneroid barometer. While these instruments worked on the same principle as the liquid barometers of the 17th century, in aneroid devices air pressure is registered by a mechanical device rather than a liquid. This idea was first proposed in 1698 by the German mathematician Gottfried Leibniz, a coinventor of calculus (see CALCULUS). However, the technology for developing such a device was not available in the 17th century, and at the time many doubted that it would ever be manufactured. By the mid-1800s advances had been made in both instrumentation and metallurgy. This enabled Lucien Vidie (1805–1866) to manufacture an aneroid barometer that used steel springs as the detection device. The instrument quickly found widespread application in both meteorology and maritime industries. There were numerous improvements in the device, mostly in the spring mechanisms and reducing the effects of temperature on the springs. However, the basic premises of the instrument are consistent with modern barometers.

The numerous variations of the barometer are indications as to the importance of the instrument to the science of the 17ᵗʰ century. Together with the thermometer (see THERMOMETER), it was now possible to quantify natural phenomena. The trend in the 17ᵗʰ century was from purely observational science toward quantitative analysis and the barometer was an important component of this change. The invention of the barometer resulted in new understandings as to the structure of the atmosphere and the ability to predict weather events. They would contribute to the discovery of laws on the behavior of gases, such as Boyle's Law. In our modern world, barometric pressure readings are common on the nightly news and form the basis of sophisticated weather-monitoring systems worldwide.

Selected Bibliography

Middleton, William E. *The History of the Barometer.* Baltimore, MD: Johns Hopkins University Press, 1964.
Spangenburg, R., and D. K. Moser. *The History of Science from the Ancient Greeks to the Scientific Revolution.* New York: Facts on File, 1993.
Wolf, A. *A History of Science, Technology and Philosophy in the 16ᵗʰ and 17ᵗʰ Centuries.* New York: Macmillan, 1968.

Binomial Theorem (1653–1676): In mathematics, a binomial is an equation that involves two variables, for example $(a + b)^n$. Like most mathematical principles, the development of a series of rules and principles to solve binomial equations received its first widespread documentation during the time of the ancient Greeks, although without question earlier cultures contributed to the Greek achievements. The Greek mathematician Euclid is considered to be responsible for organizing much of the existing mathematical work of the Greeks into one work entitled *Elements*. *Elements* contained descriptions of Greek number theory (see FERMAT'S LAST THEOREM) as well as solutions for binomial equations where the degree of the equation (n) is less than two. It is unlikely that Greek mathematicians required the use of higher-degree formulas, and while other cultures may have investigated them along theoretical lines, it would not be until the 17ᵗʰ century and its scientific advances that such a level was required. However, before serious work could begin in this area there was a need for a standardized method of mathematical symbolism for use within equations. René Descartes is credited with developing a system of notation around 1637 as part of his work on analytical geometry (see ANALYTICAL

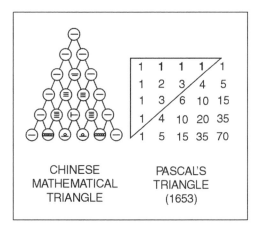

Figure 3. Mathematical triangles used to solve higher-degree equations. The Chinese triangle on the left dates from the 11th century. Pascal's triangle was used in the 17th century for the study of binomials as well as probability. The numbers represent the binomial coefficients for each term.

CHINESE MATHEMATICAL TRIANGLE

PASCAL'S TRIANGLE (1653)

GEOMETRY). The symbolism that is used in modern algebraic equations closely follows this work.

Initially the study of binomials focused on determining the coefficients of each expression in a binomial equation. For example if $(a + b)^2$ is expanded, the resulting equation is $a^2 + 2ab + b^2$. The coefficients of the expressions in the equation are 1, 2, and 1. For higher-degree equations of $(a + b)^n$ there was a need to quickly determine the coefficients of each term in the expansion. The French mathematician Blaise Pascal is often credited with developing a method of easily determining the coefficients in a binomial equation. In 1653 as part of his work on probability with Pierre de Fermat (see PROBABILITY), he developed what is commonly called a "Pascal's Triangle," although this work was not published until 1665. The operation of a Pascal's Triangle was relatively simple (see Figure 3). A grid is constructed with the first row and column consisting of the number 1. To obtain the next set of numbers one adds the numbers immediately above and to the left. For example, in the figure the 3rd number in the second line (3) is obtained by adding 1 and 2. To obtain the coefficients, a triangle is constructed whose sides are one greater than the degree of the equation $(n +1)$. For example, in the binomial $(a + b)^3$, each side would be four digits in length. The corners are then connected to form the base of the triangle. The numbers along the base represent the binomial coefficients. In our example of $(a + b)^3$, a third-degree equation, the base of the triangle has the numbers 1, 3, 3, and 1. These correspond to the coefficients for each term in the expanded equation $a^3 + 3a^2b + 3ab^2 + b^3$.

While Pascal is often credited with its development, the Pascal's Triangle actually has its origins centuries earlier. The Hindus used a similar system, called the *Meru Prastara*, around the 3rd century B.C.E. in their studies of literature. The Hindus found this triangle useful in calculating the number of different words that could be formed using a set number of letters. In the 13th century Chinese mathematicians described a mathematical triangle that had been in use since the 11th century for the purpose of solving higher-degree equations. This triangle is exceptionally similar in design to the one developed by Pascal six centuries later (see Figure 3). Furthermore, in the mid-1500s the German mathematician Michael Stifel (ca. 1486–1567) described the use of a similar triangle. However, it was Pascal who first widely used this method in his research, and thus is often credited with its development.

By the middle of the 17th century a number of mathematicians were working on a unifying binomial theorem that provided a means of solving higher-degree equations. Credit for the first significant development in this field is frequently given to the English mathematician Isaac Newton. Newton's contributions to the mathematics (see CALCULUS) and science (for example, see GRAVITATION) of the 17th century are many. In the early stages of his career Newton had developed a binomial theorem that he published in 1676. This theorem is represented by the formula:

$$(P + PQ)^{m/n} = P^{m/n} + (m/n)\,(AQ) + ((m - n)/2n)(BQ)$$
$$+ ((m - 2n)/3n)(CQ) + \ldots$$

The variables *A, B,* and *C* represent the terms of the equation, while the *m/n* value defines the coefficient for each term. Although Newton had developed the theorem, the mathematical proof was published in 1713 by the mathematician Jakob Bernoulli.

The development of a binomial theorem gave mathematicians the ability to solve higher-degree equations. It represented an evolution of mathematics from the historical basis of the ancient Greeks. However, its impact on mathematics and the science of the 17th century was overshadowed by the invention of the calculus by Newton and Gottfried Leibniz (see CALCULUS). While the binomial theorem proved useful in studies of probability and early statistics (see PROBABILITY), it was calculus that allowed scientists to mathematically explore the natural world.

Selected Bibliography

Boyer, Carl B, and Uta C. Merzbach. *A History of Mathematics.* 2nd ed. New York: John Wiley & Sons, 1989.

Cooke, Robert. *The History of Mathematics: A Brief Course.* New York: John Wiley & Sons, 1997.

Katz, Victor J. *A History of Mathematics: An Introduction.* Reading, MA: Addison-Wesley Longman, 1998.

Maistrov, L. E. *Probability Theory—A Historical Sketch.* New York: Academic Press, 1974.

Swetz, Frank J. *From Five Fingers to Infinity: A Journey through the History of Mathematics.* Chicago: Open Court Publishing, 1994.

Blood Circulation (1603–1668): Since before the time of the Greeks, blood was known to be the substance that sustained life. Ancient civilizations also recognized the heart as the mechanism by which the blood was delivered to the tissues of the body. However, the path of blood circulation in the body was less understood. The Greek philosopher Galen (ca. 129–ca. 200) developed an early model of the circulatory system involving two types of blood. He incorrectly believed that blood was manufactured in the digestive glands and then passed through the liver. Since the blood at this point contained nutrients, it was called "natural spirits." The blood then traveled to the ventricles of the heart where it was mixed with life-giving properties, the "vital spirits," before being consumed in the tissues of the body. Since Galen worked only on dead tissues, and in these tissues only the veins were observed to contain blood, he concluded that the veins alone carried the blood of the circulatory system, while the arteries carried the life-giving air. Moving the blood between the two sides of the heart presented a severe challenge to this theory, since there were no obvious direct vein connections between the sides. Galen suggested, without confirmation by experimentation, that the ventricles of the heart were linked by invisible pores, called **septum**, whose purpose was to allow the blood to move freely between the two sides. Despite the fact that there was no evidence supporting his ideas, Galen's incorrect model of the circulatory system dominated theories on blood circulation for almost 1,000 years.

The first indirect challenge to the Greek philosophy of blood circulation came in 1242 from the Arabic scholar Ibn an-Nafis (ca. 1288). Ibn an-Nafis was one of the first to describe the heart as two mechanical pumps, with one pump moving blood to the lungs, while the other moved blood to the tissues of the body. Not only was he correct in identifying the function of the chambers of the heart, but he also suggested the existence of two separate circulatory systems in the body. He was also revolutionary in explaining the role of the lungs in gas exchange. While his discoveries were significant, they had little impact on Western studies of the circulatory system as details of an-Nafis's work were not known in Western societies until the early 1900s. In the centuries immediately preceding the 17th century, an increased interest in human anatomy and physiology had led to challenges in the Galenic philosophy (see HUMAN PHYSIOLOGY). By the 1500s, notable scientists such as Andreas Vesalius (1514–1564) had demonstrated the inconsistencies in Galen's methods and reasoning. Other scientists, such as Michael Servetus (1511–1553) and Realdo Colombo (ca.1510–1559), had questioned the existence of Galen's "pores" and instead opted for the developing theory of two circulatory paths within the body. Unfortunately, Servetus's opinions on both science and religion branded him as a heretic, and eventually cost him his life. However, the previous conceptions of the circulatory system were fading, although it would not be until the 17th century, with its reliance on experimental testing of ideas, that the old theories would finally be dispelled.

In 1603, an Italian physician named Girolamo Fabrici (1537–1619) observed that veins in the legs had a series of valves along their length. Fabrici noted that these valves appeared to stop the blood from flowing downward and pooling in the feet. However, this observation meant that the blood in the veins flowed toward the heart, and not away from it as the Galenic model had suggested. While Fabrici did not openly question Galen's doctrine on the circulatory system, one of his students, William Harvey, used this observation to begin his revolution on the physiology of blood circulation.

From his studies with Fabrici, Harvey began to openly question the Galenic model. He observed that when the arteries of the body were cut, blood left the vessels in spurts, while the blood from severed veins did not display as much pressure. He also noted that the two forms of blood proposed by Galen, the "natural" and "vital" spirits, had similar characteristics, and thus might actually be the same substance. But what Harvey is best known for was his experimental approach to determining the structure of the circulatory system. Through this experimental approach Harvey demonstrated that there was a single, closed circulatory system in the body, and that the arteries, veins, and heart were all components of this system.

William Harvey first focused his attention on the function of the heart. While Galen had proposed that the heart served as a mixing station, Harvey viewed the heart as the mechanical pump of a unified system. To study the action of the heart, Harvey used cold-blooded animals, since the slowness of their circulatory systems made the examination of the process easier. This is a prime example of the use of a model system, or one that simplifies the process being examined, in the biological sciences. What Harvey noticed was that when the **atrium** of the heart contracted, it forced blood into the **ventricles**. When the ventricles contracted, the blood was forced into the arteries and to the tissues of the body. The veins then returned the blood from the tissues to the atriums. By examining the structure of the valves between the atrium and ventricles, Harvey determined that the blood in the heart moved in a single direction. Fabrici's description of the valves in the veins suggested further that blood flow was unidirectional.

William Harvey, however, was a firm supporter of experimental confirmation of observations. To verify his ideas, Harvey designed a series of experiments in which he restricted blood flow through arteries and veins and observed where the blood pooled in the blood vessel (see Figure 4). He found that when blood flow was restricted in an artery, the blood always pooled on the side of the vessel that was between the block and the heart. The opposite was true of experiments with veins. This confirmed that the vessels of the circulatory system carried blood in one direction only. This use of multiple experiments to confirm observations and hypotheses was a novel concept to the sciences, but the strength of this method was soon realized and duplicated in other investigations of the life sciences.

Harvey's experiments had firmly challenged the idea that the blood was consumed by the tissues. At the time the instruments did not exist for him to determine the link between the veins and arteries, as the microscopes of the time lacked the magnification potential (see MICROSCOPE). However, Harvey was convinced that the link did exist. Once again he designed a series of experiments to build a scientific foundation for his ideas. In one set of experiments, Harvey measured the total volume of blood in the heart. He then multiplied this value times the average pulses per minute to calculate the total volume of blood moved by the heart in one hour. Using this quantitative data, he concluded that the heart pumped a volume of fluid that was several times in excess of the total blood volume of the organism. This would not be possible if the tissues were consuming the blood. This evidence, when coupled with his studies of the structure of the heart and veins,

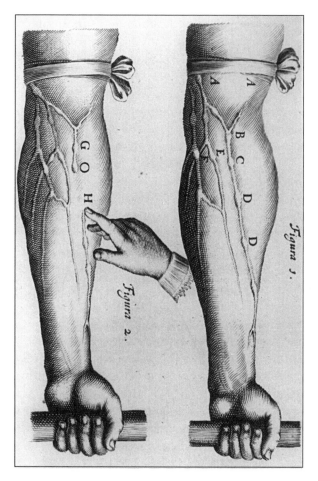

Figure 4. An illustration of William Harvey's experiment on blood circulation (1628). This experiment demonstrated the one-way flow of blood in the circulatory system. (From *Exercitato Anatomica de Motu Cordis et Sanguinis in Anamalibus,* National Library of Medicine photo collection.)

further strengthened Harvey's concept of a closed circulatory system in which the blood was recycled continuously. In 1628, Harvey published a book entitled *On the Movement of the Heart and Blood in Animals,* which outlined his conclusions and the experimental method by which they were reached. While often criticized by skeptics, this single publication revolutionized the investigation of the life sciences. Harvey's approach moved biology from theories based solely on observation and speculation to ones that were based on experimental data. In doing so he introduced the process of quantitative reasoning into the biological sciences.

As the 17th century progressed, refinements in the microscope were resulting in an increased interest in studying the microscopic nature of biological systems. In 1658, Jan Swammerdam (1637–1680), a Dutch naturalist, discovered the existence of red blood cells in the circulatory system. In 1661, Marcello Malpighi examined the structure of the circulatory system in the thin membranes of bat wings. What he noticed was a series of microscopic vessels, later called capillaries, which linked the arteries and veins. And around 1668, using a microscope with a much higher magnification potential, Anton van Leeuwenhoek observed the movement of blood cells within the capillaries. These experiments would finally confirm Harvey's theories on the structure of the circulatory system, and spell an end to the 1,000-year reign of the Galenic concepts.

The determination of the structure and function of the major components of the circulatory system would not have a great impact on the people of the 17th century. While it was not until the next century that these concepts would start to be utilized in medicine, Harvey's work did promote some preliminary interest in the study of transfusions. Around 1660 the English physician Richard Lower (1631–1691) described a transfusion of blood between two dogs. Several years later both Lower and the French physician Jean-Baptiste Denis (1625–1704) performed transfusions between an animal and man. Unfortunately, these procedures were rarely effective and thus outlawed in many countries. The medical use of transfusions would have to wait until some significant advances in understanding blood chemistry in the 20th century. However, physiological studies of blood circulation continued in the 18th century. For example, in the 18th century investigations by several scientists used developing theories on fluid dynamics to study blood flow (see HYDROMECHANICS). Stephen Hales (1677–1761) examined the pressure of blood in the body and James Keill (1673–1719) made calculations of total blood volume and rate of flow. Based on the pioneer explorations of William Harvey, these discoveries would further the development of a new method of investigation in the life sciences. As with the other sciences of the century, observations and hypotheses would have to be supported by experimental data before they would be accepted as scientific fact.

Selected Bibliography

Duffin, Jacalyn. *History of Medicine: A Scandalously Short Introduction.* Toronto: University of Toronto Press, 1999.

Spangenburg, R., and D. K. Moser. *The History of Science from the Ancient Greeks to the Scientific Revolution.* New York: Facts on File, 1993.

Botanical Classification (1623–1686): There were two major advancements in the field of **botany** during the 17ᵗʰ century. One focused on investigations into the physical structure of plants (see PLANT MORPHOLOGY), while the other concerned itself with the daunting task of botanical classification. The classification of plants, or **taxonomy**, had changed little from the time of the ancient Greeks until the period of the Renaissance. The Greeks, under Plato (ca. 427–ca. 327 B.C.E.) and Aristotle, had initially developed a classification system that focused on the soul of the organism, with plants being described as having a "vegetative soul." The first Greek botanist who focused on plant classification was Theophrastus (ca. 371–ca. 287 B.C.E.). Theophrastus classified over 550 species of plants from around the Mediterranean, but he is remembered more for the accuracy and the detail of his work. Theophrastus made a number of important contributions to the field of botanical classification. He was one of the first to classify plants by multiple characteristics. His descriptions of seed leaves, or cotyledons, and the differences between **angiosperm** (enclosed-seed) and **gymnosperm** (naked-seed) plants remained authoritative until the time of the 17ᵗʰ century. His work was augmented some centuries later by Dioscorides (40–ca. 90), a Greek physician who focused on the taxonomic classification of plants based on their medicinal properties.

During the Renaissance there was a resurgence in interest in the study of botany. However, the actual classification of plants progressed little from the science of the ancient Greeks. The majority of the work being done during the early part of the 16ᵗʰ century continued to classify plants not on their **morphological** characteristics, but on their medicinal values. However, two German contributors during this time, Otto Brunfels (1489–1534) and Leonard Fuchs (1501–1566), did add a large number of new plant species to the known database during their lifetimes. As the century progressed the accumulated number of identified plant (and animal) species required a more efficient system of organization than those that had been historically used. At the transition to the 17ᵗʰ century several attempts were being made to develop such a classification system. In 1585, the Italian botanist and physician Andrea Cesalpino (1519–1603) published *De Plantis* in which he classified plants on the basis of their flower and fruit types. While not practical for all plants, since these structures are limited to the angio-sperms only, it was an important development in plant classification and included descriptions of over 1500 species.

At about the same time, the Swiss botanist and physician Gaspard Bauhin (1560–1624) was developing a classification system that fac-tored in the natural relationships between the species of plants. In 1623 he published *Pinax Theatri Botanici (Illustrated Exposition of Plants)*, which has been recognized as a significant botanical work for several reasons. First, Bauhin had compiled descriptions of over 6000 species of plants, the cumulative total of most species known during the time. But more importantly, Bauhin introduced the concept of identifying species using binomial nomenclature (two names), although his prin-ciples were later refined by other botanists, such as Cesalpino. Under Bauhin, each species was identified by its **genus** and **species** name. This method of classification had a strong influence on the thinking of later naturalists such as John Ray (1627–1705) and Carolus Linnaeus (1707–1778). Bauhin made other contributions to 17th-century botany as well. He was a pioneer in the development of herbariums, or botanical mu-seums, where plant specimens are dried and stored in special cham-bers. Herbariums are important centers for botanical classification research and remain important components of modern botanical re-search institutions. Caspar Bauhin's brother, Jean Bauhin (1541–1613), was also a botanist who in the 17th century published detailed descrip-tions of over 5000 known species of plants.

While Gaspard Bauhin and Cesalpino had established a system, there remained a considerable amount of confusion over the definition of what constituted a species. Up until the time of the 17th century, and for some time later, it was widely believed that monsters and beasts could be formed from the interbreeding of separate species. This greatly complicated any attempts to develop a taxonomic system if each new generation had the capability of representing a distinct species. The English naturalist John Ray was one of the first to develop what is called the species concept. Ray defined a species as a group that is able to breed among themselves and produce fertile offspring. Fur-thermore, he contended that species originated by common descent and were not grown from the seeds of other species. He also suggested that some variation in a species was probably the result of **mutation**. His ideas were published between 1686 and 1704 as the important work *Historia Plantarum*. However, this was not Ray's only contribution to botany. In *Methodus Planatrum Nova (New Method of Plants)* Ray was one of the first modern botanists to make the taxonomic separation between the **monocotyledons** and **dicotyledons**. During his lifetime Ray described over 19,000 species of plants and also contributed to the taxonomic classification of insects (see ENTOMOLOGY) and mammals.

The 17th century was a time of great transition for the study of botany. By the end of the century botanists had not only described a fourfold increase in plant biodiversity, but they had also begun to develop an organization system for their data. The contributions of Ray, Bauhin, and Cesalpino greatly facilitated the description of new species using binomial nomenclature. While important, the truly significant developments in the field of botanical classification were made during the next century by the Swedish botanist Carolus Linnaeus. Linnaeus's contributions to botany were many, but he is most famous for developing a binomial system, initially applied to flowering plants, that allows organisms to be classified quickly and with a relatively high degree of accuracy. This system would eventually go on to form the basis of modern taxonomic studies; however, it has its foundations in the early botanical work of the 17th century. In 21st-century science botanical classification remains a priority among biologists. It is estimated that over 98 percent of the plant species that once inhabited this planet are now extinct. While botanists have identified over 200,000 species of plants, the actual number of species may be ten times this number. Yet many identified and unknown species are lost each year to habitat destruction and global environmental problems, mostly in the rain forest **biomes** of the planet. The rate of species loss is unequaled in modern science, and lends a new urgency to the work begun by the 17th-century botanists.

Selected Bibliography

Morton, A. G. *History of Botanical Science.* London: Academic Press, 1981.
Serafini, Anthony. *The Epic History of Biology.* New York: Plenum Press, 1993.

Boyle's Law (ca. 1661): Dating back to the time of the ancient Greeks, all matter was considered to be composed of only four elements. The Greek philosopher Empedocles (ca. 495 B.C.E.) was probably the first to suggest that all substances on Earth were derived from combinations of air, water, earth, and fire. This view of the elements was not unique to the ancient Greeks as the Indian and Chinese civilizations of the time had developed similar viewpoints. The Greek philosopher Aristotle proposed that these elements were arranged into layers, with no space, or vacuums, between the layers. The element air, for example, was located between the indiscernible layer fire and the visible elements earth and water. Aristotle also suggested that since all substances on the planet were derived from these four elements, it was possible to change materials by altering the ratios of the four elements in a process called **transmutation** (see CHEMISTRY). The Aristotelian view of the elements persisted until the time of the Scientific Revolution, when the invention of instruments to examine the physical properties of the natural world made it possible to investigate the true nature of the element air.

Many scientists of the 17th century spent time exploring the physical properties of air, including notable scientists such as Galileo. During the time of Galileo, miners had noted that it was not possible for the suction pumps to raise water to a height of more than 33 feet. This created a special problem for the mining industry as the miners often had to pump water from the lower depths of the mines. Galileo became interested in this problem, and had even suggested an experiment to test how high liquids denser than water could be pumped. However, it does not appear that he experimentally tested his ideas before his death. He was successful, however, in transferring his interest to one of his students, Evangelista Torricelli. After Galileo's death, Torricelli devised an experimental method of examining the relationship between density and the height of a column of liquid.

Torricelli and a colleague named Vincenzo Viviani selected liquid mercury for their experiment. The **density** of liquid mercury is 13.5 times that of liquid water. Torricelli designed an instrument that consisted of a mercury-filled glass tube inverted into a container also containing mercury. When the stopper was removed from the top of the tube, the level of mercury in the tube fell, but did not completely drain from the tube. In subsequent trials, the level of mercury in the tube was consistently suspended thirty inches above the level of the mercury in the container. Viviani suggested that the weight of the atmosphere on the liquid in the container was the mechanism that supported the liquid within the tube. It was observed that the weight of the atmosphere had the capability of supporting a column of water thirty-three feet tall, or a column of mercury, a denser liquid, thirty inches tall. What Torricelli had invented was a device called a barometer, which is also called a Torricelli Tube (see BAROMETER). A barometer measures the pressure of air, and while Torricelli's barometer was primitive, its principles are still used in modern devices. Furthermore, the experimental evidence that the atmosphere had a weight to it quickly dispelled the Aristotelian view of air as a weightless substance located above the element earth.

Around 1645, Otto von Guericke, a German physicist, developed the first air pump. Using this pump, von Guericke was able to remove the air from spheres to create a vacuum. By measuring the weight of the spheres before and after the experiment, von Guericke was able to make some of the first calculations as to the density of air. He also used these spheres to demonstrate the power of air pressure. In one experiment, von Guericke placed two metal hemispheres, called Magdeburg spheres, together and pumped the air from the interior space. As the pressure on the inside of the spheres dropped, the air pressure on the exterior held the spheres together, so tightly, in fact, that two teams of horses were unable to separate the spheres. When the air pressure was restored, however, the hemispheres were easily separated. The development of the vacuum pump would have a strong impact on future scientific investigations (see VACUUM PUMP).

Robert Boyle was one of the first scientists to examine the physical properties of air pressure from a mathematical perspective. Boyle was not a proponent of the Aristotelian views of the four elements, and in his book *The Skeptical Chemist* (1661), he outlined the major flaws of the four-element theory. He suggested that there were smaller components, or elements, that could be combined to form new compounds. Boyle recognized that if the atmosphere had density or weight, then it must also have a defined height. Thus, the weight of the atmosphere could not be uniform because the pressure would compress the elements of the atmosphere at the lower levels. To test this hypothesis that air may be compressed, Boyle designed an experiment using a J-shaped tube (see Figure 5). By first trapping an air pocket in the short section of the tube, and then adding a dense liquid (mercury) to the longer end of the tube, Boyle demonstrated that the size of the air pocket decreased as liquid was added. He noted that the volume of air varied in an inverse ratio to the pressure. In other words, at a constant temperature (k), the pressure of a gas (p) varies inversely as its volume (v). Mathematically, this can be expressed as

$$pv = k$$

Thus, any increase in pressure (p) must be met by a corresponding decrease in volume (v) at a constant temperature (k). This observation later became known as Boyle's Law, although the relationship was probably first discovered in independent experiments by a number of

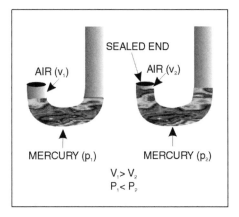

Figure 5. Robert Boyle's experiment demonstrating that the volume of a gas (v) is compressed with an increase in pressure (p). In the tube on the right, a greater amount of liquid mercury is added to the tube (p_2), compressing the volume of the air (v_2) in comparison to the tube on the left.

investigators, most notably the English scientist Robert Hooke. Research later in the century by the French physicist Edme Mariotte resulted in the same conclusions. In parts of Europe, Boyle's Law is often called Mariotte's Law. While Boyle's Law indicates the relationship between pressure and volume for a gas, there are limits to which a gas can be compressed. It was not until the next century, however, that the Swiss mathematician Daniel Bernoulli (1700–1782) first demonstrated the limits of air compression. Boyle's Law is a fundamental law of gases and became an important component of later investigations into atomic theory (see ATOMIC THEORY) and the development of compression pumps and **manometers**.

Since air compresses under pressure, logic dictates that air pressure must be greatest at sea level and lowest at higher elevations. In 1648, the French mathematician Blaise Pascal, using a barometer, measured the change in air pressure with an increase in elevation. To do this he measured the change in pressure as one traveled up a mountain. He noticed that as one progressed up the mountain the barometer indicated a corresponding decrease in pressure. Pascal calculated that air pressure dropped by a factor of ten for every twelve miles of altitude change. So while the air pressure at sea level might support a column of water thirty-three feet tall, at an altitude of twelve miles the pressure would only support a column a little over a yard in height. The altimeter of an aircraft functions using the same principles. However, instead of measuring changes in the level of a liquid, these instruments employ aneroid, or liquid-free, barometers (see BAROMETER). It was quickly noted that air pressure did not only vary with changes in altitude. In 1660, the Englishman Robert Hooke used a barometer to measure changes in air pressure during storms. Hooke noted that air pressure dropped as a storm approached. The use of barometers to measure changes in air pressure as a result of weather events resulted in the creation of a field of science called meteorology (see METE-OROLOGY).

Having entered the 17th century thinking of air as one of four unchangeable elements, the scientists of the first half of the century were instrumental in not only dispelling the Aristotelian ideas of air the element, but also identifying some of the fundamental properties and laws of air. Armed with this knowledge, scientists toward the end of the century focused on the application of these new ideas. Vacuum pumps, air pumps, and more sophisticated barometers were all the result of the early investigations of Boyle, von Guericke, and Torricelli into the properties of air pressure.

Selected Bibliography

Krebs, Robert E. *Scientific Laws, Principles and Theories: A Reference Guide*. Westport, CT: Greenwood Press, 2001.

Middleton, William E. *The History of the Barometer*. Baltimore, MD: Johns Hopkins University Press, 1964.

Taton, Rene, ed. *The Beginnings of Modern Science*. New York: Basic Books, 1958.

C

Calculating Machines (1623–1674): Historically, humans have always used calculating aids when performing mathematical operations. The first of these devices that used moving parts was most likely the abacus. The modern version of this instrument is believed to have been invented by the Chinese in the 11th century and subsequently introduced into Europe. It consisted of a series of ceramic beads that the operator physically moved for calculations (see SLIDE RULE). There were a number of instruments designed during the 17th century that were designed to assist mathematical calculations. The Scottish mathematician John Napier (1550–1617), inventor of logarithms (see LOGARITHMS), promoted the use of Napier's Bones around 1617. In the 1620s the Gunter's scale was introduced by English mathematician Edmund Gunter (1581–1626), while the mechanical slide rule (see SLIDE RULE) was invented by the English mathematician William Oughtred (1574–1660). However, each of these devices simply aided calculations; they did not truly automate the calculation process. While there is no formal definition of what constitutes a calculating machine, it is often considered to be a device in which the operator enters information but does not physically perform the calculations. These instruments frequently rely on mechanical systems to perform the desired operation.

Credit for the invention of the first generation of automatic calculating machines is often given to the French mathematician Blaise Pascal. However, early in the 20th century evidence was discovered that placed first credit for the idea with the German scientist and mathematician William Schickard (1592–1635). In a series of letters to the German astronomer Johannes Kepler (1571–1630) around 1623,

Figure 6. Blaise Pascal's 17th-century calculating machine. The device had the ability to conduct simple addition and subtraction operations.

Schickard described a device that mechanically performed addition operations. Schickard's machine was designed to work using a group of toothed shafts that the operator moved to enter numbers into the machine. While a few of these machines may have been built, it is not clear as to what extent Kepler utilized them in his work. The first machines that were produced in significant numbers, a few of which survive to this day, were invented by Pascal.

Pascal made many contributions to science and mathematics in the 17th century, including work on probability theory, hydraulics, and air pressure (see BOYLE'S LAW; HYDROMECHANICS; PROBABILITY). It is believed that his interest in calculating machines stems from a desire to assist his father, an accountant, in routine calculations. In 1642 Pascal invented a calculating machine that used a series of wheels connected to small pins (see Figure 6). The operator entered numeric data by dialing in the appropriate numbers. During the calculation a small weight was used to indicate a carry to the next decimal place. In practice the device functioned in much the same way as a modern **odometer**, as the first wheel moved from 9 to 0 it directed a change from 0 to 1 in the next decimal place. With Pascal's instrument it was possible to perform both addition and subtraction, although the process of subtraction was complicated by the fact that the wheels only moved in a single direction. Unfortunately the mechanical basis of these machines was unreliable and they remained in production for only a limited number of years.

In 1674 the German mathematician Gottfried Leibniz invented the first instrument that had the ability to perform multiplication and division operations. While Leibniz is most often recognized as the coinventor of calculus (see CALCULUS), he also contributed to the development of mathematical notation during the 17th century (see ANALYTICAL GEOMETRY). Leibniz's mechanical calculator used a rotating drum to facilitate calculations. Connected to the drum were a series of teeth whose length corresponded to a number between zero and nine. The operator entered the number on the drum, and then rotated the device so that the number that had been entered was added to the existing number in the drum. This device was also the first that allowed for multiplication and division. The rotating drum was a popular design, and although it did not find widespread use in the 17th century, it did serve as a model for 18th-century attempts to develop advanced calculating devices.

However, another of Leibniz's inventions would have a great impact on later calculating machines. Around 1679 Leibniz developed a number system based on the number two, although he did not publish his results until 1701. This system was called a binary system since all numbers were expressed as combinations of the numbers one and zero. It does not appear that Leibniz directly invented the system as part of his work on calculating machines; rather he used it in philosophical arguments on religion where one represented God and zero an absence of God. However, the discovery found more of an application in mathematics. In the common base-10 system, each place to the left of the decimal represents an increase by a factor of 10, so the number 17 represents 1 group of 10 and 7 units. In a binary system each place represents an increase by a factor of two. Thus the number 2 is represented by 10, which indicates one group of 2 and no single units. The invention of the binary system does not appear to have any real impact on the calculating machines of the 17th century as these machines were all designed around the base-10 system. However, in electronic calculating machines the binary system becomes very important. Digital electronic circuits typically operate in one of two states, "on" or "off." Thus if an electronic circuit is on, it may be used to represent a one, while an off circuit indicates a zero. The invention of electronic calculating machines in the 20th century revolves around the use of the binary system. Devices such as ENIAC, the first completely electronic computer (1946), and the invention of the handheld calculator (1971) are the result of advances in electronic circuitry that allow them to exploit the power of a binary system of calculations.

The 17th century was a time for experimentation in the field of calculating machines, although a number of them were produced and found application in many areas of society. The mechanical basis of calculations proposed by Pascal and Leibniz remained the basis for most mechanical calculators well into the 18th century. Modern calculators, and their relatives the computer, are an integral component of modern society. While primarily instruments of the 20th and 21st century, their principles of operation were established by a few mathematical pioneers and inventors of the 17th century.

Selected Bibliography

Bud, Robert, and Deborah Jean Warner, eds. *Instruments of Science.* New York: Garland Publishing, 1998.

Wolf, A. *A History of Science, Technology and Philosophy in the 16th and 17th Centuries.* New York: Macmillan, 1968.

Calculus (ca. 1684–1687): The 17th century was a time of great transition for the field of mathematics. The principles of mathematical thinking had changed little from the time of the ancient Greeks. Around the 4th century B.C.E., the Greek mathematician Euclid first documented many of the axioms, or truths, concerning geometry, the primary area of mathematical study of ancient times, and these formed the basis for most of the work in mathematics for the next 2,000 years. However, the scientific revolution that began during the Renaissance period required that a corresponding change be made in mathematical thinking to facilitate the complex calculations in the evolving fields of astronomy and physical science. The first major advance was the development of a branch of mathematics called analytical geometry (see ANALYTICAL GEOMETRY). While the study of geometry predated the ancient Greeks, in the early 17th century there developed an interest in being able to define geometric principles using algebraic equations. This required the development of a coordinate system to define the location of the geometric shape in space so that an equation could then be assigned to specific points on the shape. While many contributed to the development of coordinates, the majority of the credit is given to French philosopher and mathematician René Descartes for the invention of the Cartesian coordinate system. Descartes made another important contribution in his standardization of mathematical symbolism in algebraic equations, an important component of calculus (see ANALYTICAL GEOMETRY). Others, such as Pierre de Fermat, used analytical geometry to define the equations of complex geometric shapes such as hyperbolas and parabolas. Fermat's methods in analytical geometry would form the building blocks for the development of the calculus later in the century.

Unfortunately, many mathematical questions are not able to be solved solely by the use of geometry and algebra. In the physical world objects frequently have rates of change in relation to time. In calculus these rates of change are called *derivatives*. This can best be illustrated by examining an object being dropped from a predetermined height. The speed of the object is not a constant that can be solved using algebraic principles. Instead, the speed increases with the amount of time that the ball is falling. For these more complex analyses, it is often useful to define a dependent variable in terms of a second, or independent, variable. The result is what is called a **function**. In the early 17th century, scientists were interested in defining functions using the processes of **integration** and differentiation. Integration is an inherent aspect of geometry and involves the determination of volumes, areas, and distances. It was not new to the 17th century; the Greeks had a process called the method of exhaustion that was used by Euclid and later Archimedes. On the other hand, differentiation was a product of the development of analytical geometry and involves the finding of minima and maxima values for functions (see ANALYTICAL GEO-METERY for figure) and the tangents for curves. Before calculus was formally proposed, two different approaches were made at solving complex problems, both of which provided an important prelude to the subsequent development of calculus.

The first of these was the use of **infinitesimals**. In his investigations on the laws of planetary motion the German physicist Johannes Kepler stated that an imaginary line stretched from the center of the sun to the center of a planet will sweep the same area in a given amount of time. Thus, planets further from the sun move slower than planets closer to the sun (see KEPLER'S THEORY OF PLANETARY MO-TION). If the planets had moved in circular orbits, the calculation of these areas would have been simple. However, Kepler correctly proposed that the orbits were elliptical in shape, and thus he had to devise a method of calculating the area contained within the elliptical orbit to validate his theories. Initially using a method that he had designed to calculate the volume of a wine cask, Kepler reasoned that he could divide the shape into a large number of thin circular sections, the areas of which could easily be calculated using existing formulas for the area of the circle. If the areas of these circles were then added together, the resultant value would approximate the value for the original shape. Another method that was probably inspired by Kepler's was developed around 1635 by the Italian mathematician Bonaventura Cavalieri (1598–1647). Cavalieri was primarily interested in accurately finding the area beneath a curve, a problem that had plagued mathematics since the time of the Greeks. Cavalieri's solution was to divide the area into an infinite number of lines. These lines were then used to represent the area in a process Cavalieri called the "method of indivisibles." Unfortunately, Cavalieri's ideas were difficult to follow until they were reformed by the Frenchman Gilles Roberval (1602–1675). In modern mathematics this method is often called the Cavalieri principle and is frequently used in introductory calculus courses.

An important transition between the precalculus use of infinitesimals and indivisibles and the differentiation power of true calculus was provided by the English mathematician Isaac Barrow (1630–1677). Barrow's methods of determining the tangents to lines foreshadowed the methods that were to be developed in the calculus, and his work had an important influence on the later coinventors of calculus, Isaac Newton and Gottfried Leibniz. Barrow was one of the first to recognize that the processes of integration and differentiation are inverse properties. This would eventually lead to the formation of the fundamental theorem of calculus (see below). Others provided similar transitional work. Fermat's studies on calculating the minima and maxima values of a function, and further establishing the methods of figuring the tangent to a curve, were an important step in later work on differentiation. In addition, the English mathematician John Wallis (1616–1703) applied the method of indivisibles to a large number of problems, and also developed a new expression for *pi* based on this work. What was needed was to bring together these concepts into a single coherent principle that included a standardization of symbolism.

The unification of ideas was performed independently by Isaac Newton and Gottfried Leibniz. During the 17th century there was a considerable amount of controversy concerning which of these men was the true founder of calculus. However, it is now recognized that since a significant amount of work had already been performed earlier in the century by a number of mathematicians, and that both Newton and Leibniz had access to their findings, these two men most likely formulated what is now considered to be modern calculus independently of one another at about the same period of time. Still, calculus is a vast field, and each of these men made separate distinct contributions to the discovery.

Calculus may be divided into two branches, although the two are interconnected mathematically. The first is integral calculus and it is associated with the determination of areas under curves. It is built upon the previously mentioned concepts of infinitesimals and indivisibles. The second branch of calculus is differential calculus, which, as the name implies, addresses the calculation of rates of change. These two branches of calculus are linked by what is called the fundamental theorem of calculus, which may be expressed as:

$$\int_a^b f(x)\,dx = F(b) - F(a)$$

The notation presented in this formula is a contribution of Leibniz who first presented it in a series of publications between 1684 and 1686. The integration symbol ($\int$), as well as the term dx to symbolize differentiation, are the contributions of Leibniz. In fact, the majority of the symbolism used in modern calculus is derived from Leibniz. Leibniz also had a strong influence on two important Swiss mathematicians of the early 18th century, Jakob and Johann Bernoulli (1667–1748).

Isaac Newton had an intense interest in the development of calculus from a different perspective, that of differential calculus, or the calculus of motion. Newton was one of the first to incorporate the idea of time into his mathematical calculations. Newton considered a curve to be the fluid motion of a single point, which he called a *fluent*. Newton was interested in measuring the rate of change of the fluent, a rate that he called a *fluxion*. His study of fluxions became the basis of the differential calculus, which he is credited with discovering. First outlined in 1671, Newton did not formally present the idea until 1736. Regardless, fluxions became an important component of his research into the physical properties of the universe. The unification of this work with the powers of integration and differentiation to produce a true calculus was first presented by Newton in 1687 in his publication *Mathematical Principles of Natural Philosophy*. Newton demonstrated the usefulness of the calculus method in *Principia*, his epic work on the unifying forces of nature. In *Principia* Newton applies calculus to the study of gravitation, the laws of motion, and the movement of the planets (for example, see GRAVITATION; LAWS OF MOTION). The strength of *Principia* as a scientific masterpiece is a tribute to the mathematical principles that it contains.

The calculus developed by Newton and Leibniz toward the end of the 17th century would be a culmination of the work of many mathematicians, and not truly an invention by a single individual. Despite the success of Newton and Leibniz in formulating such a method, calculus would continue to be reformed over the next several centuries. However, the strength of the original method has endured. In our time calculus is an important component of society. The success of modern engineering from the design of bridges, buildings, and tunnels to the building of international space stations all rely on the mathematical principles of calculus and the ability it provides to analyze rates of change over time. The study of calculus is no longer confined to engineers and mathematicians; biological and chemical scientists regularly use calculus to analyze the physical properties of the microscopic world. While crucial to the creation of the physical theories of the 17th century, calculus will remain an important tool for scientists well into the future.

Selected Bibliography

Cooke, Robert. *The History of Mathematics: A Brief Course.* New York: John Wiley & Sons, 1997.

Katz, Victor J. *A History of Mathematics: An Introduction.* Reading, MA: Addison-Wesley Longman, 1998.

Kline, Morris. *Mathematical Thought from Ancient to Modern Times.* New York: Oxford University Press, 1972.

Stillwell, John. *Mathematics and Its History.* New York: Springer-Verlag, 1989.

Cells (ca. 1658–1696): Early in the 17th century the study of biology underwent a revolution following the invention of the microscope. A new discipline of science, **microbiology**, began to form as scientists began their exploration of the microscopic world. Prior to the microscope, little was known about the specifics of how organ systems functioned and organisms reproduced. Because of this, a number of misconceptions had formed regarding biological processes. For example, from the time of the ancient Greeks, animals were considered to reproduce by the process of spontaneous generation (see ANIMAL GENERATION). For centuries this idea persisted until the microscope revealed the presence of sex cells, or gametes. As the 17th century progressed, with the aide of the microscope many historical problems in biology were being answered. However, the majority of the early work with microscopes focused on the structure of plants and animals (see ENTOMOLOGY; PLANT MORPHOLOGY). This would change when in 1665 Robert Hooke published the first description of microscopic cells. Hooke was a multidisciplinary scientist who was active in a number of scientific fields, from optics to theories of fossil formation. Hooke made a number of contributions to the field of microscopy, but the one for which he is most recognized is the publication of *Micrographia* (1665). In this work Hooke brilliantly illustrated and described his observations of the individual compartments of dead cork tissue (see Figure 7). These he called *cells* since they reminded him of the small compartmentalized living quarters of monasteries. Although Hooke is credited with discovering cells, it is possible that advances in the microscope had allowed others to observe such structures at around the same time. In fact, Jan Swammerdam had proposed that the circulatory system consisted of cells as early as 1658. Regardless, it was Hooke who began the descriptive era of cell studies and the development of terminology. However, Hooke did not understand the function of these compartments in the physiology of the plant, although he did propose that they might be involved in the transportation of water. Thus a distinction needs to be made between the discovery of the first cells and the development of the cell theory. The cell theory defines the role of the cell as the fundamental unit of life. This theory was first proposed in the 19th century for plants by the German botanist Matthias Schleiden (1804–1881) and for animals by the German biologist Theodor Schwann (1810–1882). This marked the true beginning of the discipline of cell biology.

While Hooke had introduced cells to biology, the majority of the investigative work on cells during the 17th century was performed by the Dutch inventor and scientist Anton van Leeuwenhoek. Leeuwenhoek played an important role in the development of the microscope as a scientific instrument, specifically in the improvement of the magnifying systems (see MICROSCOPE). Over his fifty-year scientific career, Leeuwenhoek made a number of significant contributions to microscopy, most of which were submitted to the Royal Society of London, a prominent scientific organization of the 17th century. His observations of the microscopic world were unparalleled in the science of this time and he became a pioneer explorer into the hidden universe of microscopic life.

Unlike Hooke, Leeuwenhoek focused his attention on living cells and his discoveries often provided important verifications of developing biological theories of the 17th century. Perhaps one of the most recognized cells that Leeuwenhoek is credited with discovering is human spermatozoa. Although spermatozoa were probably first identified a few years prior to Leeuwenhoek by Louis Hamm, they initially were considered to be a disease-causing agent. In 1677, Leeuwenhoek correctly identified spermatozoa in their correct role as part of the human reproductive process. Leeuwenhoek was one of the first to describe the existence of spermatozoa in a wide variety of animals, from insects to man. During the 17th century many believed that the human body was completely preformed within the egg of the female, and that the male

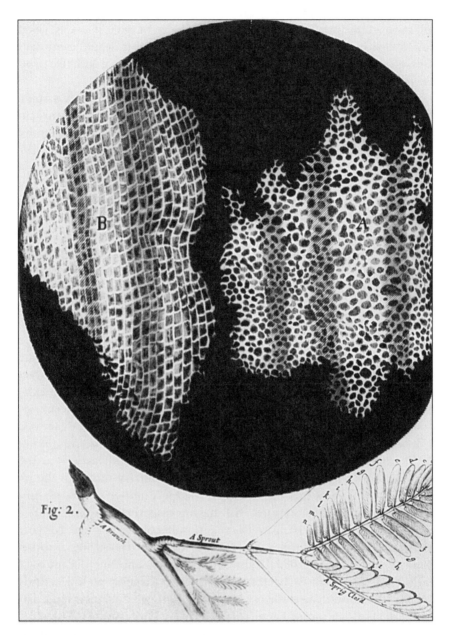

Figure 7. Robert Hooke's illustration of cork cells. (From *Micrographia*, National Library of Medicine photo collection.)

semen simply activated the growth phase. Some placed the preformed human inside the head of the sperm cell. Although Leeuwenhoek himself favored preformationist ideas, the discovery of male sex cells, when coupled with the work of other 17th-century scientists, would eventually raise questions as to the validity of preformation as a reproductive process (see ANIMAL GENERATION).

Leeuwenhoek also aided investigations into the circulatory system. In 1661 Marcello Malpighi had discovered the capillaries of the circulatory system, thus completing the circuit between the arteries and veins that had first been proposed by William Harvey earlier in the century. However, it was Leeuwenhoek who first detected red blood cells moving within these vessels, although the red blood cells themselves had first been described by Jan Swammerdam around 1658. This combination of discoveries confirmed Harvey's theories on how blood moved within the circulatory system (see BLOOD CIRCULATION).

In 1675 Leeuwenhoek made another important discovery. While examining pond water, he detected small cells moving around inside of the water. While today these single-celled organisms are called **protozoans**, Leeuwenhoek called them *animalcules* as their motion was similar to that of miniature animals. He even made early predictions as to the size of these microscopic animals, comparing their length to a grain of sand. While others had made size comparisons, the craftsmanship of the Leeuwenhoek microscopes greatly enhanced the accuracy of the observations. Furthermore, when surveying the world around him, he found an abundance of these creatures. In one experiment, he removed some of the plaque from between his teeth and examined it beneath a microscope. To his surprise there were a large number of these single-celled microscopic creatures in his mouth as well as the world around him. As he continued this preliminary survey of the invisible world, Leeuwenhoek identified a large number of microscopic cells. Leeuwenhoek is credited with having discovered the multicelled **rotifers**, the intestinal parasite *Giardia*, now becoming a major contaminant of freshwater supplies worldwide, and other free-living and parasitic protozoans. In 1683 Leeuwenhoek made another discovery of early cells, this time of bacteria. Just before the end of the century Leeuwenhoek published *Arcana Naturae* (1696), which contained a preliminary classification of bacteria based on his observations. The modern method of classifying bacteria that focuses on the shape of these primitive cells was not developed until 1773 when the Danish biologist Frederick Muller (1730–1784) first classified bacteria as bacillus (rod) and spirilla (spiral) shaped.

The work of Hooke and Leeuwenhoek in the 17th century pioneered a new field of biology and opened the doors to a realm of science that contained far more forms of life than most scientists had previously thought existed. The echoes of these discoveries persist until our time. Modern microbiologists are not only constantly finding new bacteria on our planet, some of which live in environments hostile to all other forms of life, but **exobiologists** have focused the search for extraterrestrial life on microbes. Thus, while these findings of cells in the early 17th century would not instantly change public opinion on the processes of spontaneous generation or reproduction, they would set the stage for the 19th-century development of the cell theory for plants and animals, as well as discoveries in the 20th and 21st centuries that would further define the functions of life at the cellular level.

Selected Bibliography

Serafini, Anthony. *The Epic History of Biology.* New York: Plenum Press, 1993.
Spangenburg, R., and D. K. Moser. *The History of Science from the Ancient Greeks to the Scientific Revolution.* New York: Facts on File, 1993.
Taton, Rene, ed. *The Beginnings of Modern Science.* New York: Basic Books, 1958.

Chemistry (1597–1661): The Greek philosopher Aristotle proposed that all matter was composed of one of four elements—air, fire, earth, and water. A fifth substance, ether, filled the void between the spaces of these elements. It was possible to transmute, or change, matter by altering the ratio of these four elements. It was this principle that gave rise to the early chemical science of alchemy. Alchemy was not unique to the Western world; it most likely originated in the Middle East and may also be found in the Chinese and Indian cultures as well. While most alchemists focused on transmuting metals into gold, the science also has its roots in the belief that the ability to alter chemical composition has the added benefit of prolonging life. The application of alchemy to medicinal practice is called **iatrochemistry** and had become popular in India and China by the 1300s. These practices persisted until the time of the 17th century and were not firmly refuted until several centuries after that. Until that time alchemy was often still considered a form of science. In fact, the modern term "chemistry" originated around the time of the early 17th century as a science that studied matter, while alchemy retained its focus on the transmutation of metals. Many historians consider that the study of modern chemistry was initiated as a challenge to the practice of alchemy.

The 17th century was a time of tremendous change for science, and chemistry was no exception. Like the other disciplines, the scientists of the early 17th century built on the ideas of others in an attempt to challenge the outdated Greek view of the world. In chemistry this challenge of Greek authority was directed at the Greek definition of an element as well as the practice of alchemy. Most of the chemical science of the 17th century was founded on the work of the 16th century Swiss philosopher Paracelsus (1493–1541). Paracelsus was responsible for establishing several important chemical principles and his work had a strong influence on the development of chemistry in the first half of the 17th century. Being a physician, Paracelsus was a firm iatrochemist whose interests reflected a desire to improve the chemical basis of medicine. Specifically he focused on the power of specific elements in the treatment of disease, and introduced the philosophy of purifying and quantifying chemical compounds for medical use.

At the turn of the 16th century there were a large number of iatro-chemists, and a number of attempts were being made to document their remedies and research. Notable among these was the German physician Andreas Libavius (ca. 1540–1616). Libavius is credited with developing the first chemistry textbook, *Alchemia*. First published in 1597, *Alchemia* is recognized as one of the earliest attempts to treat chemistry as a distinct scientific discipline. This highly descriptive work summarized the advances in chemistry to date. It included detailed instructions for preparing a number of chemical reagents as well as directions for conducting industrial chemical reactions such as the manufacturing of sulfuric acid. However, Libavius was not solely a chemical historian. He is also credited with the first descriptions of a number of chemical reactions that indicate his participation as a researcher. *Alchemia* also contained directions for the use of chemical instruments and details for the creation of a chemistry-based laboratory. While some question the influence of Libavius's work on science of the 17th century, as the century opened it did mark a change in scientific thought toward the discipline.

In the opening decades of the 17th century one of the prime questions facing chemical investigators was the nature of chemical reactions. There existed two possibilities. First, during a chemical reaction the reactants were transmuted in an Aristotelian manner to produce a new compound (such as lead to gold). An alternative hypothesis was that the original reactants remained intact in the new compound, although the finished product might take on distinct chemical properties. A number of early chemists were involved in this debate. One of

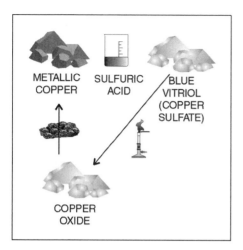

METALLIC   SULFURIC   BLUE
COPPER   ACID   VITRIOL
(COPPER
SULFATE)

COPPER
OXIDE

Figure 8. Angelus Sala's experiment with copper demonstrating that chemical elements retain their identity in chemical reactions. In this experiment copper is reacted in a series of steps with sulfuric acid, heat, and charcoal. The end result is approximately the same amount of material that began the experiment.

the first to break with Aristotle's philosophy that matter could be changed by simply altering the ratio of the four elements present within it was the French alchemist Jean Beguin (ca. 1550–ca.1620). Between 1612 and 1615 Beguin made one of the first demonstrations that compounds could be broken down into chemicals that could then be rearranged into other chemicals. Specifically, he combined antimony trisulfide with mercuric chloride to produce mercuric sulfide and antimony trichloride. This may be illustrated by the modern equation:

$$3HgCl_2 + Sb_2S_3 \rightarrow 3HgS + 2SbCl_3$$

Unfortunately, Beguin did not formally state this conclusion in his publications. However, several years later the German scientist Angelus Sala (ca.1576–1637) launched a full attack on the principles of alchemy when he demonstrated that compounds consisted of unique chemical elements that retained their identity once released from within the compound. Specifically, Sala addressed the belief that iron may be transmuted to copper, a thought that had persisted from the time of the Greeks. He demonstrated that what was actually occurring was a process that is now called **displacement**, and that there were many types of displacement reactions in nature.

From a historical perspective, what was more important was that Sala designed an experiment to conclusively prove that chemicals retain their identity in reactions, even if the final product appears to have different chemical properties. Sala turned his attention once again to copper, a popular target of alchemists (see Figure 8). Sala combined metallic copper with hot sulfuric acid to produce the compound blue vitriol (copper sulfate). When exposed to heat, copper sulfate is converted to copper oxide. By using charcoal to chemically reduce (add electrons or remove oxygen) the copper oxide, Sala was able to recover the metallic copper he started with. Furthermore, the final copper product had the same chemical properties as the original copper. When Sala weighed the copper product, he found that he had recovered the majority of the original copper, with allowable experimental losses. Thus, the final copper of the reaction was the same copper that had initiated the reaction, and it had remained unchanged during the course of the chemical process. While this would appear to be the end of alchemy, the power of alchemists and their social position in the scientific community meant that Sala's work was not widely accepted. Thus alchemy remained a factor in science until the next century.

Early chemists were making advances on many other fronts in the early decades of the century. The Belgian chemist Jan Baptista van Helmont (1580–1644) focused not on metals and alchemy, but on applying the principles of matter to the study of the atmosphere. Van Helmont is credited with the classification of gases, specifically carbon dioxide (see GASES). This played a significant role in later studies on the nature of combustion (see COMBUSTION) and respiration (see HUMAN PHYSIOLOGY). Still other chemists were applying their new knowledge in furthering studies in medicine and to industrial purposes, mostly in the development of chemical manufacturing processes.

The study of chemistry in the second half of the 17th century is connected to the development of the atomic theory, although there were advances in other areas (see ATOMIC THEORY). The prime chemist of the time was the Englishman Robert Boyle. Boyle was an opponent of Aristotelian scientific philosophy and the many significant contributions that he made to the science of the 17th century reflected this belief. He developed Boyle's Law, which defines the actions of a gas under pressure (see BOYLE'S LAW). This work on air pressure was instrumental in promoting more modern definitions of elements and atomic theory (see ATOMIC THEORY). He argued that Aristotle's four-element theory could not explain the chemical reactions of the natural world. For example, when wood is burned in a furnace it produces smoke, flames, ash, and water. While historically this had been used as evidence that the wood contained the elements fire, air, water, and earth, Boyle questioned that this was not conclusive evidence that these elements were physically present in the wood prior to combustion. Furthermore, when wood was exposed to lower levels of heat it was possible to produce oils, gases, and other compounds that were obviously not components of the four-element theory. It was this type of application of logical thinking to science that made Boyle's contributions significant.

Boyle was also a respected chemist. His 1661 publication *The Skeptical Chemist* is considered by many to be the high point of chemistry in the 17th century. Boyle made a number of significant contributions to chemistry, specifically in the areas of analytical chemistry. Rather than simply describing reactions on the basis of their interactions with heat, a process called **pyrolysis**, Boyle developed improved methods of classifying chemical reactions (see COMBUSTION) and was one of the first to quantify his experiments. He made some of the first definitions of acids and alkalines (bases) that were centered on their reactivity with common objects. While historically the classification of chemical reactions had centered on the physical descriptions of the reactants and products, Boyle instead developed a classification scheme that examined the reactivity of the compounds. This was a revolutionary step in the development of chemistry.

In a period of little over seven decades a new discipline of science had been created. However, the ancient beliefs in alchemy persisted for some time after the 17th century. Despite his many contributions to the development of chemistry, even Boyle remained convinced that alchemy was possible. Fortunately, the power of logical investigative thought would eventually prevail over the magical pseudoscience of alchemy. Chemistry now occupies a central role in modern society, with most of our 20th-century technological advances in some ways tied to chemical engineering. Credit for the evolution of this discipline originates with a small group of scientists around the time of the 20th century.

Selected Bibliography

Cobb, Cathy, and Harold Goldwhite. *Creations of Fire, Chemistry's Lively History from Alchemy to the Atomic Age.* New York: Plenum Press, 1995.

Salzberg, Hugh W. *From Caveman to Chemist—Circumstances and Achievements.* Washington, DC: American Chemical Society, 1991.

Combustion (ca. 1660–1700): Combustion, or the burning of materials, is probably one of the earliest chemical reactions viewed by man. Theories on the nature of combustion did not originate within the 17th century, although advances in the study of chemistry, air pressure, and gases did renew scientific interest in this subject around this time. Like most discussions regarding the states of matter, philosophical discussions were first documented with the ancient Greeks. The Greek philosopher Aristotle proposed the four-elements theory, which stated that all matter consisted of varying ratios of water, fire, air, and earth. The process of combustion, or burning as it was defined at the time, was considered to be the mechanism by which the element fire was released from within an object. This mechanism, with its ease of observation and compliance with the four-element theory, was logical enough to the casual observer so that there were few challenges to Aristotle's ideas until the time of the Renaissance. Followers of Aristotle, the alchemists, believed that matter could be converted by altering the ratio of elements within it. The revolution in science from the conversion of matter to the study of the properties of matter, the basis of modern chemistry, can be said to have begun with the work of Paracelsus (see CHEMISTRY). While making many contributions to the science, Paracelsus also noted that sulfur was often involved in the process of combustion. Since it was also released during combustion, Paracelsus suggested that it too should be considered an element, although this idea did not receive much support during the time. In the 17th century there began an understanding on the physical nature of air and gases (see BOYLE'S LAW; GASES) that shed new light on the existing theories of combustion.

The main problem with these early ideas on combustion is that they believed that during the process the principle elements, fire or sulfur, were being released. Yet, there was considerable evidence from metalworkers of the time that when heated, metals often gained weight. This phenomenon is called **calcination**. In modern science calcination is recognized to involve the heating of a substance to a level just below its melting point. At these temperatures the substance will often chemically react with the air, thus increasing its weight. However, during the 17th century the role of air in the processes of calcination and combustion was not well understood. Scientists of this time designed a number of important experiments in an attempt to understand the process, but did not succeed in developing a unifying theory from their results. For example, the notable scientist Robert Boyle, a major contributor to the understanding of air pressure in the 17th century (see BOYLE'S LAW), thought that during calcination it was some component of the fire or flame that contributed to the added weight of the metal. Yet in his own experiments with combustion Boyle had demonstrated that combustion could not occur in a vacuum, and thus the process required some contribution from the surrounding air. In another experiment Robert Hooke placed a combustion reaction in a container and then submerged the container under water. As the reaction proceeded, the level of air decreased, as indicated by a corresponding increase in water levels. In addition, when an animal was placed in the chamber, the animal died before the entire volume of the air was consumed. Thus Hooke proposed that air consisted of discrete substances that supported combustion and the biological process of respiration (see HUMAN PHYSIOLOGY). Unfortunately, when Boyle duplicated his experiments, this time in the absence of water, he did not detect any change in the volume of air. While Boyle was instrumental in defining laws of air pressure, he did not recognize the physical properties of gases by which the pressure of a gas will decrease to fill an available space. Thus, his experiments countered any advances in the field made by Hooke.

The English physician John Mayow (1643–1679) expanded on the Boyle-Hooke experimental plan. He demonstrated that when a candle or animal was placed in a sealed container, the air that was left over after the death of the animal or extinguishing of the candle would not support combustion. Thus, he reasoned, the air must contain two distinct components. The part of the air that supported combustion and respiration he called *nitro-aerial*. In a separate experiment he noted that combustion could be supported in a vacuum in the presence of saltpeter. Saltpeter is an **oxidizing** agent that is now frequently used in rocket propellants. To Mayow these experiments indicated the presence of discrete, invisible components in the air, some of which must be supplied by the saltpeter to support combustion. One of these components is now recognized to be oxygen, although elemental oxygen was not discovered until 1779 when it was first described by the French chemist Antoine Lavoisier (1743–1794). However, Mayow proposed that during calcination it was the nitro-aerial that was combining with the metals and responsible for the observed increase in weight. Although several others also had suggested this same idea, it was not widely accepted since the presence of these particles was difficult to confirm. Furthermore, a more reasonable theory, at least at the time, was gaining popularity.

In 1669 the English chemist Johann Joachim Becher (1635–1682) proposed a modification to Aristotle's four-element theory in which he attempted to address the problem of combustion. In place of Aristotle's elements, Becher used water, air, and three earth-like elements that were classified by their properties of combustion. Primary among these was a substance that Becher called fatty earth, or *terra pinguis.* While Becher's concept of five elements and fatty earth was primarily philosophical and did not explain the continuing problem of calcination, it did provide a basis for an experimental approach late in the century by the German chemist Georg Stahl (1660–1734). Stahl named Becher's fatty earth **phlogiston** and suggested that during the process of combustion it was this material that was being lost. In addition, he suggested that the rusting of metals to form metal oxides was also due to the loss of phlogiston and he set out to prove his ideas experimentally. He reasoned that when sulfur is placed in a combustion chamber, the phlogiston leaves the sulfur, producing sulfuric acid. Thus, if he could find a compound rich in phlogiston and heat it along with the sulfuric acid, then the sulfuric acid might reabsorb some of the phlogiston and be restored as sulfur. Stahl chose charcoal as his source of phlogiston since it had been noted previously that charcoal had the ability to restore metal oxides to their metallic form. In a series of chemical reactions Stahl combined sulfuric acid with potassium oxide (a fixing agent used to make the acid a solid). When this compound was combined with charcoal, the product had the same color and consistency as the original sulfur. In fact, the result was potassium polysulfide and not sulfur, but the science of chemistry was young and Stahl's experimental results held.

The phlogiston theory was the culmination of a century of achievements in the field of chemistry. While the theory was incorrect, it would form the basis of chemical investigations well into the 18th century. When the English chemist Henry Cavendish (1731–1810) first demonstrated the presence of hydrogen around 1766, he named it phlogiston since he believed that he had identified this unifying element. While Lavoisier later disproved the phlogiston theory in favor of one based on interaction with elemental oxygen, the 17th-century study of phlogistons resulted in several important advances for the science of the times. First, while chemistry at the beginning of the century had focused primarily on alchemy, the study of combustion marked the inauguration of the science of chemistry and study of matter. Second, even though the end result was flawed, the results were based primarily on evidence gained from experimental procedures, a significant deviation from the philosophical science of the pre-Renaissance scientists. Chemical science would progress significantly in the next century, but it had its foundations in the study of combustion in the 17th century.

Selected Bibliography

Cobb, Cathy, and Harold Goldwhite. *Creations of Fire, Chemistry's Lively History from Alchemy to the Atomic Age.* New York: Plenum Press, 1995.

Krebs, Robert E. *Scientific Development and Misconceptions through the Ages: A Reference Guide.* Westport, CT: Greenwood Press, 1999.

Salzberg, Hugh W. *From Caveman to Chemist—Circumstances and Achievements.* Washington, DC: American Chemical Society, 1991.

E

Electricity (1663–1672): As one of the more visible forces of nature, electricity has historically been of interest to scientists. However, at the start of the 17th century little was known of the nature of electrical forces. What was known focused primarily on the study of static electricity. Since static electricity typically results in the generation of a minor attractive force, such as the bending of hair toward a charged comb, electricity and magnetism were often thought of as being the same. The word "electricity" is derived from the Greek word *elektron* meaning amber. When rubbed, amber has the ability to attract small objects such as leaves in what appeared to be the same manner as a magnet.

The first scientist to make the distinction between magnetic and electrical attractive forces was the English physician William Gilbert. Gilbert is regarded as one of the earliest supporters of the scientific method. His belief that scientific findings must be supported by experimentation had a strong influence on other 17th-century scientists. The experimental science of Galileo, Kepler, and Newton, to name a few, were all guided by Gilbert's work. In the early decades of the 17th century magnetism was thought to be the unifying force in the universe. The movement of the planets and action of falling objects were all the result of magnetism. While Gilbert also supported this idea, he recognized that magnetism was primarily a property of lodestone, an iron-containing ore. However, he knew that electricity was a force present in many non-iron compounds. His findings were first published in 1600 as *De Magnete*, one of the first important publications of the century (see MAGNETISM).

Around 1663 another German scientist, Otto von Guericke, began experimenting with static electricity, although not directly at first. As was the case with Kepler, Guericke was convinced that magnetism was the unifying force responsible for the motions of the planets. In an attempt to prove his ideas he created models of the planets. What Guericke did not know was that these models contained a significant amount of sulfur. When friction was applied, the models emitted sparks and generated a weak attractive force. What Guericke had discovered was static electricity, although he incorrectly identified it as an internal force inherent to the object, similar to the magnetic properties of lodestone. These model spheres became the first electrical generators. Guericke made one additional contribution to the study of electricity. In 1672 Guericke discovered that the electrical force generated from the spheres could be used to cause the surface of a sulfur-coated ball to glow. This process is called **electroluminescence** and by the end of the century a number of other scientists, such as Dutch physicist Christiaan Huygens (1629–1695), had begun similar investigations. Guericke's publication of *Experimenta Nova* in 1672 summarized his discoveries of electricity and served as a bridge for the information into the next century.

Unfortunately, these experiments did not receive a lot of attention during the 17th century. A few years later Isaac Newton would present his ideas on gravitation (see GRAVITATION). Newton's defense of gravitation as the unifying force in planetary motion (see KEPLER'S THEORY OF PLANETARY MOTION) quickly gained acceptance with many physicists, and in a relatively short period of time interest in magnetism and electricity faded. However, Guericke's work was not without its importance. In the early 1720s the English scientist Stephen Gray (1696–1736) duplicated Guericke's experiments in static electricity. He demonstrated that this electrical force could be projected down a string of silk. He also showed that the conductance of electricity was dependent on the composition of the string; i.e., certain materials carried more electricity than others. This early work on insulators and conductance renewed interest in the study of electricity, and a number of important discoveries were made in the 18th century.

The contributions to the study of electricity in the 17th century were relatively minor in comparison to advances made in other fields of science and mathematics. However, what they did indicate was an interest in examining natural forces from an experimental basis, and thus marked the end of the Greek methods of philosophical science. While the invention of true electrical generators may be said to have occurred in the 18th century, the foundation of these discoveries is rooted in the early discoveries of Gilbert and Guericke in the 17th century.

Selected Bibliography

Spangenburg, R., and D. K. Moser. *The History of Science from the Ancient Greeks to the Scientific Revolution.* New York: Facts on File, 1993
Taton, Rene, ed. *The Beginnings of Modern Science.* New York: Basic Books, 1958.
Wolf, A. *A History of Science, Technology and Philosophy in the 16th and 17th Centuries.* New York: Macmillan, 1968.

Embryology (1604–1670): The science of embryology involves the study of how the fetus is produced and develops. In the 17th century, research in embryological science closely paralleled an ongoing debate on the subject of animal generation. Prior to the Renaissance, dating back to the time of Aristotle, there existed the belief that life may be spontaneously generated from inorganic material. However, this was proven to be incorrect in 1668 by the experimental approach of the Italian scientist Francisco Redi. Redi demonstrated that sealed containers of rotting meat material did not produce flies as was previously thought, but rather what was required was a physical depositing of eggs on the meat by a parent (see ANIMAL GENERATION). Even though support for spontaneous generation existed well into the 18th century, popular scientific support for this idea was eroding by the time of the mid-1600s. In its place there still existed two conflicting theories of how the embryo developed. These were the theories of preformation and epigenesis. The prevailing theory of preformation stated that the embryo was completely formed inside the adult, with growth involving only an enlarging of the structures over time. Some actually believed that little people were present inside the sex cells of the adults. Opposing this theory was that of epigenesis, which favored the gradual growth and development of embryonic structures inside the parent (see ANIMAL GENERATION). Even among epigenesists there was debate on what provided the nutritive substances for embryonic growth, with some contending that the menstrual blood was involved and others the semen or even urine. To support their claims, both the epigenesists and preformists conducted some early studies on embryonic development. Thus, it was this debate that resulted in some important discoveries in the field of embryology during the 17th century.

In the late 16th and early 17th centuries the majority of work on embryology was centered on studying the anatomy of the developing fetus. A large portion of this research was conducted on chicken eggs with the focus being on the role of the yolk and egg whites in the development of the embryo. As embryology as a distinct science did not exist until the 18th century, scientists from a wide variety of disciplines undertook studies of embryology during the 17th century. One of the many scientists in the 17th century who made contributions was the Italian physician Fabricus d'Aquapendente. Fabricus is most often recognized for performing some of the earliest work in comparative embryology. He is known to have dissected embryos from a wide range of organisms, including humans, rabbits, dogfish, viper, and deer, to name just a few (ca. 1604). He also developed a series of detailed drawings of embryonic development in chickens. Despite the amount of information on embryology that he had accumulated, his work was plagued with inaccuracies. He mistakenly thought that the heart of the fetus did not serve a function, and that it beat simply to preserve its own life. He also misidentified the role of sperm in the production of the embryo, which he believed never entered into the egg but instead lent its "spirit" to the egg to promote growth, an idea that had existed in some form since the time of Aristotle. However, as the 17th century progressed Aristotelian ideas, including those in embryology, began to fall under the combined weight of scientific logic and experimentation.

Although there were many contributors to embryology in the early 17th century, one of the first to implement logic and experimentation into scientific research was the English physician William Harvey. Harvey is most often recognized for his contributions to the study of blood circulation, and his publications inspired many of the scientists of the 17th century (see BLOOD CIRCULATION). Harvey also made significant contributions to the study of animal generation and embryology (see ANIMAL GENERATION). Harvey was a strong supporter of epigenesis. To test his ideas he constructed an experiment in which he monitored the development of the fetus in chicken embryos from fertilization to hatching. By fertilizing a series of chicken eggs at the same time, he was able to monitor the daily, and sometimes hourly, progress of embryonic development. He noted that the chicken started off as a single point of material that could not be differentiated from the rest, and gradually developed into a functioning fetus. While this would appear to be the final evidence against preformation, in fact many researchers of his time, including respectable scientists such as Marcello Malpighi, were claiming to have witnessed evidence in support of preformation. Harvey published his data in 1651 as *De Generatione Animalium*. It contained one of the first descriptions of the embryonic structure called a **blastoderm**, although Harvey did not understand its function. *De Generatione Animalium* disproved much of Fabricus' earlier work in embryology. For example, Harvey was one of the first to define the yolk as being involved in nutrition. In addition, he determined that the organs of the fetus were active and involved in the physiology of the organism. He observed the beating of the heart and the contraction of muscle tissues as well the reflex actions of the gastrointestinal tract. While his work did not definitively end support for preformation, the logic of his presentations provided strong support for those favoring epigenesis. However, the study of embryology during this time was not limited to animals. In 1651 the English anatomist Nathaniel Highmore (1613–1685) published *The History of Generation*. This included not only illustrations of embryonic development in plants, but also some of the earliest studies of embryology using a microscope. Unfortunately, Highmore believed that embryonic development involved a spontaneous formation of all body parts, versus the gradual development of epigenesis, at a predetermined point after fertilization.

While Harvey's work on embryology was contributing to the defense of epigenesis, others were researching the anatomy of the reproductive system organs. Harvey was primarily a mechanist, but later in the 17th century the French scientists Pierre Gassendi and René Descartes attempted to apply the developing theories of atomic theory to embryological studies (see ATOMIC THEORY). Descartes published several treatises on embryonic development in which he proposed that the organs of the body originated by the interaction of smaller parts. The motion of these parts in the formation of larger organs was governed almost exclusively by the laws of physics and chemistry. Other researchers explored the reproductive system from a more anatomological perspective. Many considered the ovaries to be homologous to the testes of a man. Nicholas Stensen (1638–1696), also known as Nicolaus Steno, demonstrated that the human female ovary was homologous to the ovaries found in many **oviparous** animals. Since the ovaries of these animals were known to house eggs, the conclusion could be drawn that human female ovaries had a similar structure. Another continuous debate among the early embryologists was the role of the umbilical cord in the development of the fetus. Around 1667 Walter Needham correctly determined that the umbilical cord was involved in supplying nutrients to the developing fetus, and was not a respiratory structure.

By the end of the 17th century the microscope had become a popular instrument among biologists and was a logical instrument of choice for embryologists. The microscopist Marcello Malpighi performed a number of the early studies of embryology using a microscope. Using Harvey's data on chicken embryo development, Malpighi was able to trace the progress of the embryo hour by hour. Malpighi made a number of important contributions, including the first detailed description of the blastoderm. In addition, he detected early signs of blood vessels within the embryo. His detailed drawings augmented the existing documentation of embryonic development. Unfortunately, Malpighi was a supporter of preformation, and even believed that he had witnessed miniature human forms inside embryonic cells. Malpighi was not the only preformist who documented the existence of miniature human forms. Theodore Kerckring published a similar statement in 1670. Despite the fact that he was the first to witness the cleavage of the egg in a frog, the microscopist Jan Swammerdam remained a supporter of the preformation theory. The scientific reputation of these individuals did much to prevent counterevidence against preformation from being recognized as factual.

In many ways, the study of embryology left the 17th century in the same state of confusion as it entered. Scientific investigations had not yet progressed to the point of being able to eliminate the debate on how a human embryo develops. In the next century more detailed investigation on the growth of the embryo, and specifically the differentiation of tissues, would be conducted. The concept of epigenesis, as supported experimentally by Harvey in the 17th century, was redefined as the differentiation of material followed by growth and became the prevalent theory of the time. In addition, while there were relatively few chemical investigations of embryology in the 17th century, mostly due to the fact that this discipline of science was still in its infancy (see CHEMISTRY), chemical investigations of embryology occurred more frequently in the 18th century. While there would still be some setbacks in these later studies, the study of embryology made strong progress during the 1700s. This was due, in part, to a good foundation in observational embryology established during the 17th century. In 21st-century science, the study of embryology is linked closely with the topics of genetics and genetic counseling. Modern embryologists have studied the embryonic development of a large number of organisms, and frequently use this information to construct the evolutionary history of a species. Furthermore, in the 21st century the use of ultrasound and screening of the amniotic fluids has made it possible not only to predict the precise stage of development of a human embryo, but also to determine whether genetic problems may exist in the developing fetus. The recent completion of the Human Genome Project will greatly enhance the genetic study of embryology. By many predictions, it may soon be possible to genetically engineer embryos to remove certain birth defects and undesirable traits, a massive step forward from the initial studies of the 17th century.

Selected Bibliography

Gasking, Elizabeth. *Investigations into Generation: 1651–1828*. Baltimore, MD: Johns Hopkins University Press, 1967.

Keller, Eve. "Embryonic Individuals: The Rhetoric of Seventeenth-century Embryology and the Construction of Early-modern Identity." *Eighteenth-Century Studies* 33, no. 3 (2000): 321–348.

Needham, Joseph. *A History of Embryology*. New York: Abelard-Shuman, 1959.

Pinto-Correia, Clara. *The Ovary of Eve: Egg and Sperm and Preformation*. Chicago: University of Chicago Press, 1997.

Entomology (1602–1710): Entomology is the branch of the biological sciences that involves the study of insects. As the origin of insects predates the history of man, our species has always interacted with insects, specifically in the areas of medicine and agriculture. The scientific study and classification of insects is considered to have begun during the time of the early Greeks. As with many other aspects of Greek science, the philosopher Aristotle served as a major force in the development of this discipline. During his time Aristotle identified some 500 species of animals, many of which were insects. While performing these classifications, Aristotle divided the insects into two major groups based on whether they are wingless or winged. He then subdivided the groups by the structure of their mouthparts (teeth, proboscis, etc.). He also made some of the first contributions toward the understanding of insect anatomy and morphology. Unfortunately, Aristotle's studies also led him to falsely conclude that insects arose by spontaneous generation. He frequently proposed that insects generated from the inanimate materials in which the adults of the species were found, for example, the generation of flies from manure (see ANIMAL GENERATION). This misconception persisted for centuries and would become one of the leading reasons for the study of entomology during the 17th century.

The science of entomology received a considerable amount of attention during the 17th century. Primarily this was due to improvements in scientific instrumentation, most notably the invention of the microscope. Furthermore, insects proved to be a useful system for the study of many aspects of the biological sciences. In the early years of the 17th century the Italian naturalist Ulysse Aldrovandi (1522–1605) conducted an extensive review of **zoological** literature in order to construct an encyclopedia on the animal kingdom. The ninth volume of the series summarized the known information on insects. This work was one of the first modern attempts at a large-scale classification of the insects, what is now called systematical entomology. As with many other of these early attempts at classification, there existed obvious errors in the methods by which many of the classifications were established. For example, bats and birds were classified together owing to the presence of wings. In addition, many **invertebrates, mollusks,** and various worms and **echinoderms** were lumped together as insects. However, Aldrovandi did propose a new classification system that first separated insects by their terrestrial and aquatic habitats and then by the physical presence and number of wings. While most of Aldrovandi's work was centered on the area around Italy, similar studies were being performed in other parts of Europe. The English naturalist Thomas Moufet (1550–ca. 1604) posthumously published a similar study in 1634. This work is frequently criticized for containing a large number of errors, for example, the belief that the queen honeybee was actually a male. However, together with the work by Aldrovandi it served as the reference for much of the entomological work during the remainder of the 17th century.

The concept of the spontaneous generation of insects, as believed by Aristotle and many others up to the time of the 17th century, was finally disputed by the work of the Italian scientist Francesco Redi. A popular belief until the 17th century was that fly maggots spontaneously generated from rotting meat. In 1668, Redi designed a series of experiments to disprove this idea. He prepared two sets of jars, one of which was sealed while the other was left exposed to the air. After an elapsed period of time Redi discovered that only the exposed jars had generated maggots. However, the possibility existed that fresh air was required for the maggots to spontaneously generate. In a second experiment Redi sealed one set of jars with gauze to allow the passage of fresh air. As with the previous experiments, only the unsealed exposed jars developed maggots. Not only did this experiment represent a significant advance in the study of animal generation (see ANIMAL GENERATION), but it was also one of the first documented examples of using controlled experiments to test the contribution of a single variable to an observation (see SCIENTIFIC REASONING). Redi's work was later confirmed by the experiments of Antonio Vallisnieri, who became one of the primary entomologists in Europe in the 18th century. Additional verification was obtained from Dutch microbiologist Anton van Leeuwenhoek, who detected the eggs using his improved microscopes (see MICROSCOPE). Thus, it was the study of insects that finally disproved the long-standing belief in spontaneous generation.

A number of scientists during the 17th century developed an interest in the anatomy of insects. The Italian scientist Marcello Malpighi, an important contributor on the anatomy of the human circulatory system (see BLOOD CIRCULATION), made a number of important discoveries regarding the internal structure of insects. Notably, he discovered the excretory system of insects, now called Malpighian tubules in his honor. He is also recognized as being one of the prominent investigators of anatomy of the silkworm *Bombyx*, an insect of economic importance throughout the world. The invention of the microscope and its improvement throughout the 17th century opened new opportunities for the study of insect anatomy. Leeuwenhoek, the inventor of some of the best microscopes of the 17th century, was one of these investigators. With the aid of these instruments Leeuwenhoek discovered that the eyes of insects were actually compound and consisted of thousands of tiny facets. Leeuwenhoek's work served as an inspiration for the English naturalist John Ray. Toward the end of the 17th century Ray based his insect classification system on the work of Leeuwenhoek. Ray is recognized as an expert taxonomist, whose methods of classification were based not only on the morphology of the organism, as was historically the case, but also on the biology of the species. Ray believed that organisms could be classified on the basis of "common descent," or their relationship to a common ancestor. He applied this reasoning not only to the classification of insects, as is evident in his publication *Historia Insectorum* (1710), but also in his work on the classification of plants (see BOTANICAL CLASSIFICATION). His ideas contributed to the development of the **Theory of Natural Selection** by the English naturalists Alfred Wallace (1823–1913) and Charles Darwin (1809–1882) in the 19th century.

One of the most important contributors to the science of entomology in the late 17th century was the Dutch scientist Jan Swammerdam. Swammerdam's contributions to entomological science were numerous. He developed a strong interest in the study of insect **meta-morphosis** and classified insects as to the extent of their change. He recognized that insects, such as the butterfly, undergo a complete or holometabolic metamorphosis that consists of a defined larval stage. Other insects, such as the grasshopper, conduct a partial or hemi-metabolic metamorphosis in which the insect resembles the adult but undergoes a series of molts to reach adult size. Some insects lack a metamorphic stage in their development. These are called ametabolic and they simply grow in size without change until they reach the adult size. An example is the silverfish. To some extent this still serves as the basis of the modern classification of insects. In addition, as part of his interest in the process of metamorphosis Swammerdam helped define the adult, larval, and egg stages associated with insect development. His research was aided by his ingenuity and ability to invent the precision scientific instruments needed for work with small insects. Using these inventions he was able to identify many previously unknown structures of insects, including the venom glands and reproductive systems of bees and the anatomy of exceptionally small insects such as lice and aphids. His detailed descriptions of spiders, scorpions, and worms (then frequently considered to be insects) remained accurate well into the next century.

As with many other scientific disciplines, the modern study of entomology can be said to have its roots in the science of the 17th century. Although frequently not considered as an independent field of study during the 1600s, the study of entomology was a cornerstone of biological investigations during this important scientific century. The study of entomology in the 17th century had little impact on the culture or people of the times as the advances made did not bring an end to insect-borne disease or agricultural problems. However, 17th-century entomology represents an important advancement for the science of biology. Not only were new scientific instruments developed and tested on the study of insects, but the work being conducted was used as a confirmation of many emerging theories by biologists. These studies helped finalize the demise of Aristotelian thought and in doing so opened the door toward more experimental-based studies in later centuries.

Selected Bibliography

Serafini, Anthony. *The Epic History of Biology.* New York: Plenum Press, 1993.

Smith, Ray F., Thomas E. Mittler, and Carroll N. Smith, eds. *History of Entomology.* Palo Alto, CA: Annual Reviews, 1973.

Taton, Rene, ed. *The Beginnings of Modern Science.* New York: Basic Books, 1958.

F

Fermat's Last Theorem (ca. 1630): Pierre de Fermat was a French mathematician who played an important role in the mathematical revolution of the 17th century. Fermat's contributions to mathematics were many. His discoveries frequently closely paralleled the work of his countryman René Descartes. For example, Fermat and Descartes independently developed a coordinate system for the study of geometric figures. The discovery of coordinate geometry gave 17th-century mathematicians the ability to manipulate geometric shapes using algebraic equations. This was the beginning of the study of analytical geometry, or the defining of geometric shapes using algebraic principles, a branch of mathematics in which Fermat played a key role. He established algebraic formulas for many geometric shapes, including hyperbolas and parabolas. He also made significant contributions to the study of curves that included finding the minimum and maximum values of the curves and solving for the area under a curve (see ANALYTICAL GEOMETRY). In cooperation with the French mathematician Blaise Pascal, Fermat established early theories on the study of probability, although this work was directed initially at predicting games of chance and not scientific research (see PROBABILITY). Despite his many achievements in mathematics, Fermat is most often recognized for his work on number theories and for proposing a mathematical theorem that took over 300 years to prove.

In mathematics, the study of number theory involves defining the properties of whole **integers**. The study of the theory of numbers was not unique to the 17th century, as its history most likely dates back to the time of Pythagoras. The Greek mathematician Euclid presented one of the first comprehensive descriptions of ancient number theories, although it is widely regarded that this was a compilation of the work of earlier mathematicians. This work, entitled *Elements,* contained important definitions of what constituted a number and a prime number. Despite being published in the 3rd century B.C.E., *Elements* served as the cornerstone of mathematical thought until the time of the Renaissance. However, work on number theories was not limited to Euclid. The Roman (Syrian) mathematician Nicomachus (ca. 60–ca. 120) published an important work, *Introduction to Arithmetic,* which contained some early descriptions of proportions and geometric number theory. Several centuries later the Roman (Egyptian) mathematician Diophantus (ca. 200–ca. 284) published *Arithmetic.* The topic of this book was primarily algebra but it contained many important discussions of mathematical symbolism that were needed for the study of number theory. *Artihmetic* is known to have had a strong influence on Fermat's later work. Through the Dark Ages of Europe, Greek mathematics was kept alive in the Arabic countries by mathematicians who added to the understanding of prime numbers with their own work.

Fermat's interest in number theory initially involved the study of prime numbers for the identification of perfect numbers. A perfect number is a number that is a sum of its factors. The Greeks had recognized that if $2^n - 1$ is a prime number, then $2^{n-1}(2^n - 1)$ represented a perfect number. The Greeks had identified four perfect numbers (6, 28, 496, and 8,128), but were limited in their studies by the fact that there were relatively few values of n for which $2^n - 1$ was also prime. Fermat first developed a method of identifying any prime numbers for which $2^n - 1$ was also prime. Although called Mersenne primes after the French mathematician Marin Mersenne (1588–1648), a colleague of Fermat, these were in fact the result of Fermat's work. These numbers are the basis of what is now called Fermat's Little Theorem, which includes detailed descriptions of the derivation of prime numbers. This theorem states that the expression $a^p - a$ is divisible by p if p is a prime number. As was typical of Fermat's work, the mathematical proof of this theorem was not presented until later in the century by the German mathematician Gottfried Leibniz. Fermat was also instrumental in the development of what is called the process of infinite descent. This is a form of mathematical reasoning in the study of number theory that defines conditions under which certain positive integers can't exist. It was this type of reasoning that served as the basis for Fermat's most famous mathematical achievement.

Fermat's Last Theorem, sometimes called the Great Theorem, stated that for the equation

$$x^n + y^n = z^n$$

there is no solution if $n > 2$ and x, y, z, and n are all positive integers. It is known that Fermat began the formation of this theorem while studying the areas of right triangles using the process of infinite descent. He proved that it is not possible for the area of a right triangle to be a square. While Fermat is responsible for stating the theorem that bears his name, it is unclear as to whether he actually proved it mathematically. The only records of Fermat's work in this area are some minor notes entered into the margin of his copy of Diophantus's *Arithmetic,* along with the inscription: "I have discovered a truly remarkable proof which this margin is too small to contain."

Since that time, countless mathematicians have attempted to develop the mathematical proof of this theorem. It has been estimated that between 1908 and 1912 over 1,000 incorrect proofs of this theorem were presented. However, over time some progress was made. By the mid-1800s proof of the theorem had been divided into two main classes that focused on whether the terms x, y, and z were divisible by n. The invention of computers and their widespread use in mathematical calculations in the 20th century accelerated the proof of Fermat's Last Theorem for most values of n less than 4.0×10^6. In 1993 the English mathematician Andrew Wiles (1953–) presented a proof of the theorem as a by-product of his studies on another long-standing mathematical problem, the Taniyama-Shimure Conjecture. Initially there existed some problems with his proof, but with the cooperation of the English mathematician Richard Taylor (1962–), Wiles was able to derive a proof in 1995. This complex proof is now widely recognized as finalizing this centuries-old mathematical problem. Unfortunately, despite the fame of his work, Fermat's Last Theorem had a minimal impact on the mathematics of 17th-century Europe. One reason for this was Fermat's reluctance to publish his scientific findings, as may also be viewed in his work on analytical geometry, an area that he certainly developed before Descartes (see ANALYTICAL GEOMETRY). Without publication, few of the other mathematicians during this time period were aware of his work. While seemingly not important from a 17th-century perspective, Fermat's Last Theorem has had a tremendous impact on mathematical science over the ages. Not only have some of the brightest minds in math attempted to produce a proof for the theorem, but it has attracted the attention of multitudes of amateur mathematicians. Although there were other more significant discoveries in mathematics during the 17th century, for example the development of calculus (see CALCULUS), the longevity of Fermat's Last Theorem is often used as a representation of the advance of number theory and general mathematical knowledge during this century.

Selected Bibliography

Cooke, Robert. *The History of Mathematics: A Brief Course.* New York: John Wiley & Sons, 1997.

Katz, Victor J. *A History of Mathematics: An Introduction.* Reading, MA: Addison-Wesley Longman, 1998.

Kline, Morris. *Mathematical Thought from Ancient to Modern Times.* New York: Oxford University Press, 1972.

Swetz, Frank J. *From Five Fingers to Infinity: A Journey through the History of Mathematics.* Chicago: Open Court Publishing, 1994.

Fossils (1665–1691): Fossils are the ancient remains of once-living creatures that have become petrified, or converted, to stone over a period of millions of years. While there are several methods by which a fossil may be created, the most common involves the rapid burial of a living organism by **sedimentary** material. In modern science the process of fossil formation is recognized to be a relatively rare event as not all organisms form fossils. If the living creature possessed a shell, skeleton, or other solid components, then the rapid burial by silt or other sedimentary material prohibits bacterial decay. Over time, the hard parts of the creature's anatomy may become mineralized, thus forming a fossil. In some cases an imprint of the organism, sometimes including the soft tissues, is left on mud or silt. These may later also form fossils if they are covered by additional sediments and form new rock. However, over the history of science many alternate ideas developed on the origin of fossil remains. Furthermore, as fossilized remains may be found throughout the globe, it is highly likely that ancient scholars and scientists also debated the origins and nature of these items. Some considered fossils to be the remains of once-living creatures, while others proposed that fossils were actually nothing more than stones that had been carved by the action of water and wind. This debate would continue into the 17th century and play an important part on the evolution of geology as a science (see GEOLOGY).

During the Renaissance the **organic**, or living, nature of fossils was supported by a number of scientists such as the French engineer Bernard Palissy (ca. 1510–ca. 1589) and the Italian architect Leonardo da Vinci (1452–1519). However, they were not the first to come to this conclusion. It is known that the ancient Chinese recognized fossils as being the remains of living creatures centuries before the same realization in Western science. During the 17th century a number of other scientists continued to accumulate evidence supporting this idea. One of these was the English scientist Robert Hooke, who made a number of observations regarding fossils. Around 1665 he suggested that petrified wood had been formed from a natural event involving sedimentation, although he also believed that this event was the biblical flood. The Dutch scientist Nicolaus Steno is often presented as the first to propose a modern theory for the formation of fossils. Steno made a number of contributions to the developing field of geology (see GEOLOGY), and is widely considered to be the first real geologist. He examined shark teeth and noted that the fossilized remains often closely resembled those from living specimens. Furthermore, he noted that fossils found in different rock layers were often identical. These two observations almost completely ruled out the possibility that fossils were carved from rocks by the action of wind and water alone.

One of the primary obstacles to the study of fossils prior to the Renaissance was associated with religious and philosophical disagreement as to the age of the earth. Many considered the earth to be just over 6,000 years old, far too young for fossilized remains to have been generated. Early scientists who investigated fossils had to overcome this belief, thus slowing their progress. Many believed that the fossils were the result of the Great Deluge as presented in the Bible. The force of the flood was considered to be enough to force the fossils into the various layers of solid rock in which they were discovered. While difficult to prove scientifically, these ideas did result in the development of some important concepts during the 17th century, most importantly the involvement of sedimentation.

In his research Steno recognized that there were two different types of rock, primary rocks that lacked fossils and secondary rocks that contained the fossil record. Furthermore, by examining the rocks in which the fossils were found it was possible to determine the habitat of the fossilized creatures. By examining the types of fossils and the rocks in which they were found Steno concluded that many fossils were formed by the process of marine sedimentation and were not of terrestrial origin. As primary rocks were often found lower in the earth than secondary rocks, Steno predicated that these rock layers differed in age. This led to the development of what is now called the principle of superposition, or the fact that older rock layers are found on the bottom and become progressively younger as one moves toward the surface (see GEOLOGY). Although this theory was correct, Steno believed in a relatively young Earth and that the rock layers were formed during the biblical flood. More accurate estimates on the age of the planet would take several centuries to develop.

A continuing problem of the fossil record is that it contained evidence of creatures that were no longer in existence. This presented a special challenge for Western scientists with a religious belief in a single creation event. Two prominent scientists of the 17th century were among the first to address this problem. Robert Hooke concluded that since the fossil record contained evidence of species that were not missing from the planet, it was logical to conclude that these species may have been destroyed by some natural event. Furthermore, he suggested that if species may have been destroyed, then other species may have developed over time and not have been a part of a single creation event. Both Hooke and the German mathematician Gottfried Leibniz suggested that these older species could have been transformed into organisms that were now present, an interesting prelude to later 19th-century discussions on the process of evolution.

Scientific investigation into fossils in the 17th century was not limited to debate on their origin. Many extensive collections of fossils were begun during this time, some of which persist until modern times. Since an understanding of the fossil record often requires access to a large collection of items in order to make an accurate comparison, these collections set the stage for fossil work into the next century. In more modern times, fossils have played an important part not only in the development of geological principles, but also in the study of evolution. Debates on the interpretation and accuracy of the fossil record were responsible for many of the evolutionary theories being developed up to the 19th century and remain an important component of present-day evolutionary studies. From a 17th-century perspective, the interest in fossils began a period in which scientists started to question the age of the earth. This required a greater understanding of the forces shaping the planet, which resulted in the birth of geology as a scientific discipline.

Selected Bibliography

Gohau, Gabriel. *A History of Geology*. New Brunswick, NJ: Rutgers University Press, 1990.

Ronan, Colin. *Science: Its History and Development among the World's Cultures*. New York: Facts on File, 1982.

Taton, Rene, ed. *The Beginnings of Modern Science*. New York: Basic Books, 1958.

G

Gases (ca. 1620–1670): To the ancient Greeks, air was one of the four fundamental elements from which all matter was derived. As such, there existed very little scientific interest in breaking down air into individual gases. However, in the 17th century the development of the scientific method challenged the historical basis of Greek science. In the early part of the century a number of scientists were redefining the Greek definition of an element. During this time there were a number of discoveries and inventions that revolved around air and the atmosphere. The invention of the vacuum pump and the barometer, coupled with explorations in air pressure and combustion and the development of chemistry as a scientific discipline, all heralded a renewed scientific interest in the study of air (see BAROMETER; BOYLE'S LAW; CHEMISTRY; COMBUSTION; VACUUM PUMP). Although the majority of the work in the study of individual gases would be accomplished in the next century, a number of important contributions were made during the 17th century.

The term *gas* originated in the 17th century with the work of the Belgian chemist Jan Baptista van Helmont. Owing to religious beliefs, Helmont rejected the Greek four-element theory of Aristotle. He contended that since fire is not mentioned in Genesis, it should not be considered an element. Furthermore, he believed that air was inert and did not contribute to chemical reactions. Thus, all matter must be formed from some form of water. To test his hypothesis he repeated an experiment first proposed by Nicolas of Cusa (1401–1464) in the 15th century. He placed a willow tree in a pot containing 200 pounds of soil. For the next several years Helmont provided water to the tree but did not make any additions to the soil. At the end of the experiment he measured the weight of the tree and found that while the tree had gained 164 pounds in weight, there had not been an appreciable change in the weight of the soil. Helmont contended that the change in the weight of the tree must be due to the added weight of the water, and thus it is water that is the prime element of matter. There were a number of obvious problems with this experiment, prime among them the fact that the change in the weight of the tree was actually the result of the fixation of carbon dioxide gas during the photosynthetic process. From a historical perspective, this experiment had some important consequences. First, it was one of the first to attempt to quantify scientific measurements, but what was more important was that it established the background for Helmont's later work with gases.

Helmont's work on gases was the result of his attempts to prove that water was the fundamental element of life. It was widely known that during a chemical reaction air was often released, frequently in the form of bubbles. In Aristotle's four-element theory the release of air was not a problem as air was one of the four prime elements of all matter. This presented a serious conflict for Helmont and he devoted a considerable amount of time to proving that what was escaping during these reactions was not air, but some other compound. He proposed that this compound was some form of modified water, which he called *gas*, a term most likely derived from the Greek word for chaos. Others, such as Robert Boyle, had proposed that this compound was actually modified air. Regardless, it was becoming apparent that what had historically been considered an element may actually consist of multiple components.

Helmont designed a number of experiments in which he conducted chemical reactions and attempted to collect the escaping gas for classification. While he had only moderate success in capturing these gases, mostly because of inadequate equipment, he did describe the gases based upon their odor, color, and taste. He is credited with the first descriptions of a number of gases. He generated carbon dioxide, the contributing factor to his willow tree experiment, from fermenting grapes and combustion reactions with charcoal. By burning sulfur, one of Paracelsus's primary elements (see CHEMISTRY), he obtained sulfur dioxide. He also isolated methane by distilling organic material. Helmont was a firm believer in the atomic nature of matter, and believed that by subjecting his gases to intense cold he could restore them to a liquid state (see ATOMIC THEORY). Unfortunately this remained purely theoretical as the technology did not exist during the time to liquefy gases in this manner. Still, his scientific principles were sound, even if the final theory was flawed.

Helmont was not the only contributor to the study of gases during the century. The English scientist John Mayow proposed that air consisted of two components, one of which directly supported combustion and respiration (see COMBUSTION). This component he called *nitroaerial*. In the next century the French chemist Antoine Lavoisier demonstrated that this compound was actually oxygen gas. Robert Boyle also identified many properties of air (see BOYLE'S LAW) and recognized that air consisted of discrete components. He published a work called *On the Springiness of Air and Its Effects* in which he drew many of the same conclusions as Mayow.

At the start of the 17th century the study of chemistry was in its infancy. The study of gases was a result of this new interest in understanding the nature of matter. While the identification of many of the properties of gases and air would have to wait for later technological advances in the field of chemistry, these initial studies in the 17th century played an important role in the science of the times. While this work on gases may seem minor in comparison to other discoveries of the century, gravitation and planetary motion for example, it was directly connected to an understanding of combustion and air pressure, which were more significant discoveries of the century. As the science of chemistry was founded in the 17th century, the study of gases may be considered to be an important contributor in the evolution of this discipline.

Selected Bibliography

Cobb, Cathy, and Harold Goldwhite. *Creations of Fire, Chemistry's Lively History from Alchemy to the Atomic Age.* New York: Plenum Press, 1995.
Salzberg, Hugh W. *From Caveman to Chemist–Circumstances and Achievements.* Washington, DC: American Chemical Society, 1991.

Geology (1644–1669): Civilizations have interacted with the Earth since the beginning of time. Agriculture, raw materials, resources, and natural disasters have all shaped the history of our species. For this reason mankind has historically attempted to explain both the origins of the Earth and its natural cycles in terms that could easily be understood by ancient cultures. Furthermore, each culture tended to have its own unique interpretation on the nature of the Earth. For example, the ancient Greek philosophers Eratosthenes (ca. 276–ca. 194 B.C.E.) and Strabo (ca. 63–ca. 23 B.C.E.) developed several theories as to what may have caused fossils of marine organisms to be deposited great distances from the shores of modern seas. As with much of the work performed by the Greeks, their discussions relied little upon experimentation and facts, and more upon observation and philosophy only. The birth of geology as a science is considered to have its roots in the 17th century. During this time scientists first attempted to attach scientific principles to the study of the Earth, and rely less on philosophy, myths, or superstitions. While there was not a single unifying theory on geology developed during this century, there were a number of advances and discoveries made that set the stage for investigations well into the next two centuries.

Initially, interest in geology focused on predicting the age and origin of the Earth. In Western cultures this tended to place these studies in direct conflict with religious doctrines, which in turn had the effect of slowing scientific progress. Many scientists prior to the 17th century had developed theories on the origin and structure of the Earth. In 1644 René Descartes, a French philosopher and scientist, published *Principles of Philosophy*. In this work Descartes compared the Earth to the sun and suggested that the Earth still contained a molten core surrounded by metallic regions. Descartes used his model to make early predictions on the cause of many geological phenomena, including the formation of mountains and oceans. This was coupled with some early predictions as to the age of the Earth. Around 1650 the Irish archbishop James Ussher (1581–1656), using the Bible as a **genealogical** reference to count backward through the generations of kings in the Old Testament, calculated that the Earth was formed around 4004 B.C.E. and thus was less than 6,000 years old. This prediction would have a severe impact on the studies of geology for the next several centuries as geologists and other scientists were forced to develop theories that allowed for natural events to occur in what is now considered a very brief geological interval.

The origin of mountains attracted a considerable amount of interest during the 17th century. Descartes thought that mountains originated from the collapse of the surrounding terrain, while others considered mountains to be the remnants of the Great Flood of the Bible. As with the Greeks, these theories were almost completely theoretical. However, a few, such as the French geologist Pierre Perrault (ca. 1611–1680) and the English naturalist John Ray, had begun to integrate science into their work and viewed mountains as integral components in the movement of water (see HYDROLOGIC CYCLE) and in the formation of natural ecosystems.

The greatest contributor to the development of geology as a science in the 17th century was the Dutch scientist Nicolaus Steno. As with others in the 17th century, Steno became involved in geological studies as a result of an interest in fossils (see FOSSILS). Steno divided rock layers into two main categories that were identified on the basis of the fossil record that they contained. In this classification primary rocks lack a fossil record, while secondary rocks contain some evidence of fossils. Those rocks that contained fossils were further divided into smaller categories based on where they were formed. By closely examining the rock layers and the evidence of living material that it contained, Steno recognized that he could identify the environment in which the rock layer was formed, be it terrestrial or marine. Several centuries later this became the foundation for the definition of many geological formations.

As a result of this work, Steno developed a theory called the law of superposition. Fundamentally what this law states is that the oldest layers of rock should be located deepest within the Earth and as one progressed toward the surface the age of the rock layers should decrease. Furthermore, the oldest layers were composed of hard, rocky substrates, while the younger layers were formed by sedimentation events. The law went further to explain that since the younger layers were formed by the process of sedimentation, and this sedimentation was in the form of silt from the mouths of rivers, the deposition of the sediment should occur evenly over the ocean floor. As a result, the rock layers that were formed from this sediment should possess boundaries that were horizontal to the ocean floor, with the boundaries between the layers being roughly parallel to one another. Modern geologists call this process the law of original horizontality. Once again the major challenge to the theory was the existence of mountains, which by observation lacked horizontal structure. But on closer examination Steno determined that the rock layers within the mountains had become tilted and still exhibited evidence of a once-horizontal structure. This suggested that the rock layers had been moved over time. Although Steno incorrectly assumed that this was due to a collapse of the area in a manner supported by Descartes, rather than uplifting as is now recognized by geologists, it was an important first step in the realization that the surface of the Earth is very dynamic. Modern geology recognizes that many forces, including plate tectonics, continental drift, volcanoes, and even the weather, combine to shape the surface of the planet.

However, even to Steno it was obvious that not all rock layers remained in a horizontal configuration indefinitely. It was possible to find areas in which older and younger rocks were intermingled in complex patterns. By studying the area of Tuscany, Italy, Steno derived an explanation for this phenomenon. He suggested that in areas such as this there had been a series of geological cycles. These cycles had consisted of repeated sedimentation events followed by undermining of the layers by the action of water and the uplifting of layers, which Steno thought to actually be the result of collapse. This served to jumble the rock layers from their normal horizontal positions. He once again distinguished between the age of the rocks layers, and thus their order of formation, by the type of rock (sandy or rocky) that was present. It is unclear whether Steno may have used fossils in this dating process. Today, however, modern geologists use these same principles in their dating of rock layers and their reconstruction of the geological history of a given area.

The 17th century was an important time for the study of geology. Not only were important advances made in the study of basic geological principles, but scientists were also beginning to understand the nature of fossils and cycling of water (see FOSSILS; HYDROLOGIC CYCLE). The 17th century was the time of the birth of geology as a scientific discipline, and as such it quickly entered into the scientific revolution occurring in the Western world. Although the advances noted above had little impact for the people living in the 17th century, they did serve as the foundation for more important and influential studies done in the 18th and 19th centuries. During these times man would develop an appreciation for the true age of the Earth and the complexity of its geological structure. New disciplines, such as paleontology and archeology, would develop while existing science such as biology would quickly incorporate the information being provided by this new branch of science that originated during the 17th century.

Selected Bibliography

Gohau, Gabriel. *A History of Geology.* New Brunswick, NJ: Rutgers University Press, 1990.

Taton, Rene, ed. *The Beginnings of Modern Science.* New York: Basic Books, 1958.

Wolf, A. *A History of Science, Technology and Philosophy in the 16th and 17th Centuries.* New York: Macmillan Company, 1968.

Gravitation (ca. 1684–1687): Without doubt one of the most significant figures in the history of science was the English physicist and mathematician Isaac Newton. Long regarded as the "Great Synthesizer" of scientific concepts and theories, he held a key role in the evolution of 17th-century science. Like many of the scientists of the times, Newton was an interdisciplinary researcher who contributed to a wide range of scientific topics. From calculus and optics to the origin of comets (see CALCULUS; HALLEY'S COMET; OPTICS), the majority of the investigations into the physical laws of the universe during the 17th century were influenced by his discoveries. While many came before him, it was Newton who late in the 17th century formulated many of the unifying theories of physics. Perhaps his greatest scientific achievement was the development of the unifying theory of gravitation. His contributions to this one area summarized decades of research and provided the basis of scientific thought for centuries to come.

Gravity was not technically a discovery of the 17th century. The effects of gravity on a falling object have obviously been known since the beginning of recorded time. However, throughout history the *nature* of gravitational forces has been poorly understood. For example, the Greek philosopher Aristotle attempted to explain gravitational pull as a component of his four-element theory (see CHEMISTRY). To Aristotle, all matter was composed of four elements (fire, air, water, and earth), and objects fell to the ground in order to return to their "natural" state. However, the observed motion of a projectile complicated Aristotle's theory and remained a subject that was not adequately addressed until the work of Galileo (see LAWS OF MOTION). In the 16th century, the lack of understanding of gravity hindered the acceptance of Copernicus' model of the universe (see KEPLER'S THEORY OF PLANETARY MOTION). For if the movement of the heavens was due to the rotation of the Earth, it was difficult to understand what kept objects from flying off into space. In the 17th century gravitation was often confused with magnetism, even though William Gilbert had made some early distinctions between these forces during his research (see MAGNETISM). Even renowned scientists such as Johannes Kepler believed that magnetism was the governing force behind planetary motion (see KEPLER'S THEORY OF PLANETARY MOTION). Galileo recognized gravity as the governing force behind the actions of falling bodies (see LAW OF FALLING BODIES), but he did not comprehend the universal nature of this force.

Popular legend has it that Newton had a revelation regarding gravity after an apple fell on his head. While this is probably exaggerated,

Newton was interested in the force that directed moving and falling objects, although identifying the properties of that force would take Newton a significantly longer period of time than legend portrays. Newton's derivation of his ideas on universal gravitation, or the effects of gravity on all matter, began during his early investigations on the laws of motion (see LAWS OF MOTION). It appears that Newton became interested in the physical properties of motion from a number of different perspectives, one of them being a study of Kepler's Law of Planetary Motion. Kepler's Third Law states that the squares of the period of the orbits of two planets are proportional to the cubes of their mean distances from the sun (see KEPLER'S THEORY OF PLANETARY MOTION). To explain this mathematically Newton developed what is called the inverse-square law. Basically, this states that the strength of a force is inversely proportional to the square of the distance of the object from the source. Although others had previously suggested that such a relationship existed, Newton utilized, and expanded upon, the inverse-square law as a component of his much larger work on motion. However, Newton did not publish the results of his findings until approached by the English astronomer Edmund Halley, who was researching the motion of comets. Halley believed that he had evidence that comets moved in elliptical orbits, but he lacked the mathematics to explain this motion (see HALLEY'S COMET). After visiting Newton and learning of his work on the inverse-square law and laws of motion, Halley recognized that Newton had the solution to the motion of comets and other heavenly bodies. Halley not only convinced Newton to publish his results, but he also provided financing for the work. In 1684 Newton published *On the Motion of Bodies in Orbit*, frequently called by its Latin title, *De Motu*. This work provided the foundation for all his later work on gravitation (see LAWS OF MOTION) and eventually it evolved into his prime work, *Principia*.

From the work of previous scientists, namely Galileo and Kepler, Newton derived three laws of motion. The first law related to **inertia**, or the theory that an object at rest will remain at rest unless acted upon by an outside force. The second law stated that the change in the motion of an object, in both its magnitude and direction of movement, was proportional to the strength and direction of the force causing the change. Newton's third law stated that for every action there is an equal and opposite reaction. Now all Newton needed to do was to identify the nature of the unifying force that governed the physical laws of the universe. Newton initially experimented with the motion of pendulums. He noted that quantity of matter in an object was proportional to the weight of the object. Furthermore, it was not the type of matter that determined the motion of the pendulum, but the quantity of matter. This was significant in that it not only set the stage for Newton's examination of celestial motion, but the studies also recognized that the force involved was common to all types of matter.

While not initially interested in the motion of heavenly objects, Newton found that when he applied his laws of motion to Kepler's Third Law of Planetary Motion, the force that kept the planets in their orbits decreased with distance according to the inverse-square law. Since the Galilean satellites of Jupiter behaved in the same manner as the planets, there must be some common denominator to this motion (see JUPITER). To test his ideas, Newton turned his attention to the motion of the moon. From Galileo's study of projectile motion and inertia (see LAWS OF MOTION), Newton recognized that there must be a force exerted from the Earth that inhibited the moon from traveling in a straight line. Also, according to his own third law of motion, there must be a second force that prohibited the force projected from the Earth from dragging the moon into the Earth. This second force was **centripetal force** and was described in Newton's second law of motion as the force opposing **centrifugal** motion (see LAWS OF MOTION). Unlike others who supported magnetism, Newton thought that the attractive force was gravity and he set out to prove it mathematically. It was known that the distance from the Earth to the moon is approximately sixty times the radius of the Earth. Newton recognized that the acceleration of the moon was $1/3600^{th}$ of that found on the Earth. Newton then calculated that the force that "bent" the orbit of the moon into its elliptical shape was the result of an inward acceleration of 2.7×10^{-3} meters per second per second. An object falling on the Earth from a resting position accelerates at a rate equal to 60^2 times this amount. Since the value for the moon is equal to $(1/60)^2$ of the Earth value, Newton recognized this as being the same as the inverse-square law. When Newton compared the values for the inward acceleration of the moon to the acceleration of an object on Earth, it was apparent that a definite relationship existed. It was obvious to Newton this was the same force, the strength of which was being influenced by the distance between the two objects. This force of course was gravity. Furthermore, Newton understood that the mass of the objects must influence the effect of gravity. Earlier in the century Galileo had determined that the acceleration of a falling object was uniform and independent of the mass of the object (see LAW OF FALLING BODIES). Using this information, Newton derived the formula:

$$F = \frac{GM_1 M_2}{d^2}$$

In other words, the gravitational force (F) acting on an object is related to the mass of the objects (M_1, M_2) times a gravitational constant (G) divided by the square of the distance between the two objects (d^2). The value of the gravitational constant (G) was determined in 1797 by the English physicist Henry Cavendish.

It should be noted that not all of the credit for investigations into the theory of gravitation belongs to Newton. The English scientists Robert Hooke and Christopher Wren (1632–1723) had independently developed an inverse-square law. Hooke had even suggested that the motion of the moon was dependent on gravity. However, it was Newton who, in the spirit of 17th-century science and the reliance on mathematics and experimentation, provided the mathematical proof of gravitation and established it as a universal property. It should also be noted that universal gravitation is a more complex phenomenon than Newton first proposed. Universal gravitation is easy to understand from a terrestrial perspective because the gravitational force of the Earth dominates all outside sources. However, not all astronomical observations comply with Newton's theory, since from a universal perspective gravity is a weak force that only attracts objects and never repels them. The understanding of these deviations has played an important role in science. For example, in the 19th century astronomers noted that the orbital path of Uranus contained irregularities that could not be explained as gravitational interactions with its existing neighboring planets. By recognizing that this was due to a gravitational influence of some yet unidentified body and charting the pattern of these irregularities, it was possible in the 19th century (1846) to discover the planet Neptune. In the 20th century the German-American physicist Albert Einstein (1879–1955) performed a similar study on the planet Mercury. He noted that the behavior of Mercury could not be explained simply from gravitational interactions. To accommodate this problem, Einstein was forced to make some major alterations to Newton's work. This resulted in the development of the Theory of Relativity, the modern version of Newton's work on universal gravitation.

Newton published his findings in 1687 as *Philosophiae Naturalis Principia Mathematica*, more commonly called the *Principia*. In this monumental work, Newton finally provided the proof of Kepler's Laws of Planetary Motion with respect to universal gravitation. Newton also made one of the initial distinctions between the mass of an object, or its resistance to acceleration, and the force of the gravitational pull, or what is commonly called weight. Although Newton prepared his proofs in *Principia* using the more common geometrical methods of the time (see ANALYTICAL GEOMETRY), he also introduced the calculus as a method of analyzing rates of change (see CALCULUS). *Principia* was not simply a work on celestial mechanics, it also contained unifying ideas on the properties of light (see OPTICS). However, it was the theory of universal gravitation that set *Principia* apart from other scientific publications of the time. With one work Newton had verified the scientific achievements of 17th-century scientists, including Galileo, Kepler, and Halley. The gravitational principles set forth in *Principia* provided the background information necessary for understanding the motions of planets, comets, and falling objects, as well as the force responsible for the generation of tides (see OCEAN TIDES). In the history of science there have been few works that have had such a tremendous and lasting impact on scientific thought. *Principia* represents the culmination of a century of scientific achievement. In a space of eight decades during the 17th century scientists had ventured from discovering some initial proofs that the Earth revolves around the sun, to identifying the unifying force that structures the heavens.

Selected Bibliography

Cajorie, Florian. *A History of Physics*. New York: Dover Publications, 1962.
Christianson, Gale E. *In the Presence of the Creator*. New York: Macmillan, 1984.
Hall, A. Rupert. *From Galileo to Newton*. New York: Dover Publications, 1981.
Newton, Isaac. *The Principia.*. Translated by J. Bernard Cohen and Anne Whitman. Berkeley: University of California Press, 1999.
Spangenburg, R., and D. K. Moser. *The History of Science from the Ancient Greeks to the Scientific Revolution*. New York: Facts on File, 1993.

H

Halley's Comet (1668–1687): By the 17th century the development of scientific reasoning, new mathematical skills, and instruments dedicated to scientific investigations would forever change the way that people viewed nature. For astronomers, this century would generate scientific investigations examining the place of comets in the heavens. The results of these studies served as a confirmation of several theories on the physical laws of the universe. While the majority of this work was being performed in the new scientific observatories and laboratories of Europe, the bulk of the observational data on comets had been done centuries before by ancient civilizations.

The sudden appearance of a comet in the night sky had many meanings in ancient cultures. Unlike the brief life span of a meteor entering the atmosphere, the sighting of a comet often persisted for days or even weeks. Each evening the comet appeared to continue its leisurely movement across the heavens, with a pattern of motion that was independent of the sun, moon, or planets. To many cultures, the appearance of comets was viewed as an omen, and for this reason many early cultures kept detailed records on the appearance of these heavenly visitors. Available documents from the Sumerian and Akkadian cultures make reference to comet-like phenomena, and some of these records date back to around 4000 B.C.E. The Aztec, Mayan, and Indian cultures also record the appearance of comets throughout their history. However, the most abundant and accurate ancient observations of comets belong to the Chinese. The ancient Chinese dynasties recorded the appearance of over 600 comets, some of them dating from 2315 B.C.E. The Chinese first recognized Halley's comet during its 240

Table 1
Recorded Observations of Halley's Comet

May 240 B.C.E.	March 1066*
November 164 B.C.E.	April 1145
August 87 B.C.E.	September 1222
October 12 B.C.E.	October 1301
January 66	November 1378
March 141	June 1456*
May 218	August 1531*
April 295	October 1607*
February 374	September 1682*
June 451	March 1759*
September 530	November 1835
March 607	April 1910
October 684	February 1986
May 760	??–2061
February 837	

*Dates of the visits used by Halley in his calculations.

B.C.E. visit (see Table 1), although there are some records that they may have observed it as far back as 1615 B.C.E.

For the ancient Greeks celestial objects in the heavens were considered to be flawless and perfect in shape. Furthermore, the stars, planets, sun, and moon all resided on transparent crystalline spheres that circled the Earth and guided the nightly display of lights in the sky. The random appearance of a comet, with a motion that was often independent of the other heavenly objects and an irregular shape, presented a severe challenge to this idea. Aristotle contended that comets were not celestial objects, but were parts of the Earth that, when raised into the heavens, interacted with the element fire to produce the comets. Like many of Aristotle's views, this concept of comets as atmospheric phenomena would persist for centuries.

By the 15th and 16th centuries, European observers were taking an interest in comets. Some, such as the Italian Girolamo Fracastoro (ca. 1478–1553) and the German Peter Apian (1501–1552), noted that the tails of comets always point away from the sun, a fact that had originally been described by the Chinese 700 years prior. In the 16th century the Danish scientist Tycho Brahe (1546–1601) made an attempt to measure the distance to a comet. Brahe was a meticulous scientist who made exceptionally detailed observations on the paths of the planets. In the centuries prior to Brahe, the Ptolemaic system of an Earth-centered universe had accumulated a significant number of errors. Brahe collected his data in an attempt to correct these discrepancies and restore an accurate method of predicting the motions of the planets and stars. One of these problems was related to the study of comets. To determine which of the crystalline spheres comets resided on, he used a process called parallax analysis. In parallax analysis, an object is viewed from two observation points and its change in position in relation to the background is measured. Objects that are close to the observation points will appear to have a greater amount of movement in relation to the background than those that are far away. This can easily be demonstrated by placing an object several meters away. If you stare at the object and then alternately close your left and right eyes, without changing your position in relation to the object, the object will appear to jump in relation to the background. But if you increase your distance between yourself and the object to 100 meters, then the same experiment will result in less of a shift in the position of the object in relation to the background. When applied to the comet of 1577, Brahe noted that the orientation of the comet to the background of the night sky did not change much when the comet was viewed from observation points some distance apart. Since a similar effect occurred with the moon, it could be concluded that the comet was not located in the atmosphere, but rather resided somewhere past the orbit of the moon. Although the concept would take some time to be accepted, eventually Brahe's work ended the idea of comets as atmospheric phenomena.

The 17th century was an important time for astronomical observations. The beginning of the century had witnessed the birth of the telescope as a scientific instrument (see TELESCOPE). Shortly thereafter, Galileo's discovery of the moons of Jupiter effectively ended the Ptolemaic model of the universe (see JUPITER). The findings of these early astronomers generated an intense interest in studying celestial objects, including comets. In 1668 the Polish/German astronomer Johannes Hevelius (1611–1687) published a book entitled *Cometographia* in which he provided detailed descriptions of the various tail formations of comets observed from 1577 to 1652 (see Figure 9). Significant advances were also being made by astronomers such as Johannes Kepler on the laws of planetary motion (see KEPLER'S THEORY OF PLANETARY MOTION). Kepler's calculations that the orbits of the planets were not perfectly circular, but rather elliptical in shape, had a substantial impact on the work of comet investigators later in

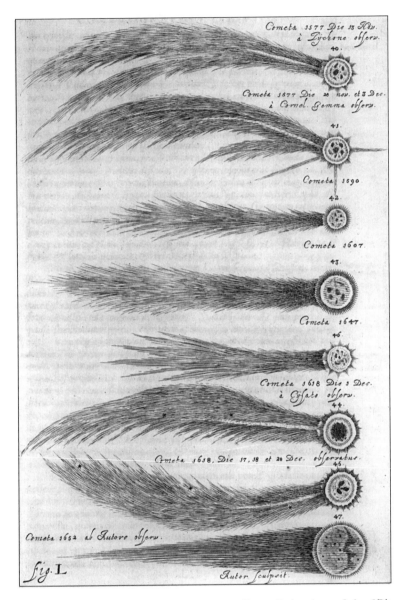

Figure 9. A description of known comet tails until the time of the 17th century as provided by the German astronomer Johannes Hevelius. (From *Cometographia* [1668], Library of Congress photo collection.)

the century. However, Kepler himself did not apply these elliptical orbits to comets. Instead he suggested that these transient objects moved in a straight line through the solar system.

These events would have an influence on an individual named Edmund Halley. Halley transformed comet science from a strictly observational basis to a form that incorporated new discoveries in the fields of planetary motion and mathematics. As an astronomer, Halley had already become recognized at a young age for his production of star maps of the Southern Hemisphere (see STAR ATLASES). Halley's interest in comets appears to have been sparked by a comet that he observed in 1680. This comet was visible in November of that year and then disappeared behind the sun. However, in December another comet was observed emerging from behind the sun. While the prevailing idea was that comets traveled in a straight line, Halley began to question whether this observation was the result of two comets, or a single comet that was in orbit around the sun. The appearance of an even brighter comet in 1682 cemented Halley's interest in studying comets. While many consider Halley to have discovered comets, especially the comet of 1682 that now bears his name, the fact is that Halley did not discover a single comet. The honor of discovering Halley's comet should be bestowed on the Chinese, who first detected it centuries earlier. Instead of discovering comets, Halley was one of the first to analyze the historical data on comets in an attempt to learn more about the orbital paths of these objects.

Problems in science often resemble a puzzle. Individual discoveries are similar to pieces of the puzzle, each making a contribution to the solution. Halley, a gifted astronomer and scientist, belonged to a rare group of scientists who were capable of pulling the pieces together to find a solution. In the study of comets, several events earlier in the 17th century had provided important background information. Early in the century Galileo had proposed that objects move freely in space, in contrast to riding on crystalline spheres. Then came the work of Kepler and his laws of planetary motion and the elliptical orbit of the planets. While Kepler did not apply his results to comets, Halley was convinced that the same force that kept the planets in their orbits would apply to all celestial objects, including comets. But Halley could not identify the force of nature that made this possible. During Halley's time, a number of scientists theorized that the force responsible for the motion of the planets was somehow directed outward from the sun. The strength of this force was predicted to have an inverse-square relationship. As the distance from the sun increases, the strength decreases as an inverse of the square. This is represented by the formula $1/r^2$, where r equals the distance. Thus an object that moves twice the distance from the sun would have one-quarter of the force directed at it. Halley understood this law, but he could not identify the force or determine how it could be applied to his study of comets.

There was, however, an individual who had the answer for Halley. Around this time the English physicist Isaac Newton was spending considerable amounts of his time developing his laws on universal gravitation (see GRAVITATION) and motion (see LAWS OF MOTION). Halley sought out Newton for assistance and after learning of Newton's studies of gravitation, Halley convinced him to publish his work. The result, released in 1687, was Newton's *Principia*, a colossal work that combined many discoveries of the early 17th century. Included were Newton's descriptions of gravitational force and its effects on the orbits of planets. This was exactly the force that Halley needed to explain the orbits of comets. Halley next turned to the recorded observations of comets from between 1337 and 1698. He found twenty-four comets during that period that had enough information available for analysis. Using the historical information, Halley calculated the perihelion, or closest point to the sun, as well as the angle of the comet's orbit in relation to the orbit of the Earth (called the **inclination**) and the orbital path of the comet around the sun. This was an exceptionally difficult task in the 17th century as no computers or calculating aids were yet available, although Halley did have access to the use of logarithms, invented earlier by the Scottish mathematician John Napier (see LOGARITHMS).

After his calculations were complete, Halley observed that most comets were random; that is, they did not originate from the same location, follow the same orbits, or possess the same inclination in their orbits. However, there was an exception. The comets of 1531, 1607, and 1682 all possessed the same characteristics. Halley questioned whether this was a single comet with an approximate 76-year orbit or multiple comets originating from the same location. Not believing that this was a chance event, Halley looked further back in the historical records. He noted that a comet had been detected 75 years earlier in 1456, and another 390 years earlier (78 years × 5) in 1066. The differences in the period of the orbits initially presented some problems for Halley. However, he reasoned that the gravitational laws established by Newton for the sun and planets should also apply to interactions between comets and planets. Thus the mass of Saturn and Jupiter should both influence the orbit of the comet. Halley then predicted that the comet would return in 1758, and although he did not live to witness it, the comet that now bears his name returned as predicted (see Table 1).

Although comets themselves were not discovered in the 17th century, until this time they had been viewed as unpredictable objects whose presence was often considered an evil omen. The mystique surrounding comets exists in some form until this day, with some groups retaining the belief that the appearance of a comet is a sign that the world is about to end. The work of Halley, built on the foundations of Galileo, Kepler, and Newton, provided evidence that comets were not transient stellar objects, but rather companions to the other occupants of the solar system. While Halley's predictions on the return of a comet probably did not significantly impact the lives of the people in the 17th century, his investigations directly influenced the work of Isaac Newton. Newton's unifying theories of gravitation and motion were first applied to Halley's investigations of comets, and it appears that Newton's *Principia* probably would have remained unpublished for some time longer had it not been for Halley's persistence. Thus, while the comet itself may not have been a significant finding, the results of the collaborative effort between Halley and Newton would have a tremendous impact on the science of the next century and beyond.

Selected Bibliography

Evans, James. *The History and Practice of Ancient Astronomy.* New York: Oxford University Press, 1998.

Moore, Patrick. *The Great Astronomical Revolution: 1543–1687 and the Space Age Epilogue.* Concord, MA: Albion Publishing, 1994.

Sagan, Carl, and Ann Druyan. *Comet.* New York: Random House, 1985.

Whipple, Fred L. *The Mystery of Comets.* Washington, DC: Smithsonian Institution Press, 1985.

Heliocentrism (1609–1687): At the start of the 17th century there existed two major theories on the physical layout of the universe. The prevailing thought of the time was called the Ptolemaic system, first proposed by the Greek philosopher Ptolemy in the second century. The major premise of this theory was that the sun, moon, planets, and stars all were in orbit around the Earth. This concept of an Earth-centered, or geocentric, universe persisted for almost fourteen centuries and became one of the longest-lasting scientific theories in history. Opposite of the Ptolemaic system was the idea that the Earth, moon, and planets all revolved around the sun. This heliocentric model is often called Copernicanism, after the Polish astronomer Nicolaus Copernicus (1473–1543), who presented the concept in 1543, although it had its roots much earlier in history. While neither of these originated within the 17th century, the wealth of data provided by the advancements in science during this century would forever close the debate between these two conflicting theories.

To early observers the motions of the stars, sun, and moon gave the indication that these objects revolved around the Earth. Even the planets, with their more erratic motions, followed a similar pattern across the night sky. Thus when constructing the first models of the universe, it was logical to designate the Earth as the central object around which all other objects revolved. While this geocentric model did not originate with the ancient Greeks, they are considered to be the first to develop mathematical models for an Earth-centered, or geocentric, universe. This started with the Greek philosophers Thales (624–526 B.C.E.) and Anaximander (610–ca. 546 B.C.E.), who envisioned the universe as a series of geometrically spaced spheres surrounding the Earth. The application of mathematical properties to the spheres would begin with Plato. The mathematical basis of the spheres was further explored by the Greek mathematician Eudoxus of Cnidus (ca. 400–ca. 350 B.C.E.) In the model proposed by Eudoxus, the objects in the heavens resided on twenty-seven shells or spheres, each of which guided the motion of a celestial object. While to Eudoxus the multiple shells were merely mathematical representations of the laws governing the movement of the objects, Aristotle later proposed that the spheres were physical objects. Furthermore, each object in the heavens was guided by the action of one or more spheres. Some celestial objects, such as the wandering planets, were guided by complex interactions between the spheres. The complexity of this model would continue to develop under the influence of the Greek astronomer Hipparchus. Hipparchus uncovered variations in the movement of the planets that could not be explained by Aristotle's idea of perfect concentric circles. Rather than develop a new theory, Hipparchus introduced the concept of an **epicycle**. This meant that the planets made small circular motions during their revolution around the Earth (see KEPLER'S THEORY OF PLANETARY MOTION). The size of the epicycle could be varied to compensate for the irregularities of the observations. While this gave enough fluctuation in the orbits to explain the observations, it would greatly complicate the formation of a unifying geocentric model for almost 200 years.

During the second century the Greek scientist Ptolemy constructed a model consisting of a complex system of interacting spheres and epicycles that controlled the motions of the planets, sun, moon, and stars. Due to the complexity of planetary motion, each planet was governed by the actions of multiple spheres and epicycles. To Ptolemy the entire model was a mathematical creation, but in the following centuries the spheres themselves would take on physical properties, as was the case with Aristotle. Despite its flaws and complexity, the Ptolemaic model would remain basically unchallenged for almost fourteen centuries.

In the centuries following the development of the Ptolemaic model, the idea of a geocentric universe had gained considerable popularity with the doctrine of the dominant Catholic Church. The consequences of challenging this ruling body greatly stifled the creation of competing models. Yet over time a number of problems began to manifest themselves within the existing system. The model had become inaccurate in its prediction of astronomical events, often missing important dates by days. Attempts were made to correct the problem by increasing the influence of the epicycles. However, these attempts further complicated an already complex model. By the 16th century the accumulated corrections and adjustments to the Ptolemaic system had made it almost unmanageable. The navigation of the long sea routes to the Americas and the Far East required accurate astronomical tables. This directed a reevaluation of the current model. The result would be two competing models, one that kept the basic principles of the Ptolemaic system, and one that completely reshaped the structure of the universe.

Opposite the idea of the **geocentrism** was the belief in a heliocentric model. While this heliocentric system had its origins during the time of the ancient Greeks—it had been proposed by the Greek astronomer Aristarchus (ca. 310–ca. 230 B.C.E.)—it did not gain popularity until the work of Copernicus in the 16th century. Copernicus proposed that the errors of the Ptolemaic system could best be minimized by centering the universe on the sun. While the new theory, **heliocentrism**, was not devoid of problems, since Copernicus still used circular orbits for the planets (see KEPLER'S THEORY OF PLANETARY MOTION), it did correct many existing problems, namely the **retrograde motion** of the planets. However, the theory was not widely accepted at the time. The lack of a unifying force created some real conflicts with the Copernican model. In a geocentric system if an object is thrown it simply falls downward toward the center of the universe, the Earth. In a heliocentric system, centered on the sun, there was no explanation for why objects fell downward. Similarly, if the Earth revolved around the sun, one could question what kept objects from flying off of the planet (see GRAVITATION).

While the existing Copernican model could not answer these questions, the Danish astronomer Tycho Brahe proposed another solution. Tycho recognized the strengths and weaknesses of the Ptolemaic and Copernican theories and proposed a synthesis of the two. First, Tycho would keep the Ptolemaic premise of an Earth-centered universe. This corrected the problem of gravity but still could not explain the retrograde motion of the planets. For this second problem, Tycho borrowed Copernican ideas and placed the planets in orbit around the sun. The sun, in turn, revolved around the Earth. This elaborate hybrid system, supported by the wealth of astronomical data accumulated by Tycho himself, would provide an interim solution to the existing problems of the Ptolemaic system without removing the Earth from its dominant position at the center of the universe. However, this idea would be short-lived.

By the beginning of the 17th century advances in astronomy would spell an end to scientific support of geocentrism. The invention of the telescope in the early 1600s (see TELESCOPE) would allow astronomers to notice that the heavens were full of stars, far too many to be controlled by crystalline spheres. Galileo's discovery of the moons of Jupiter in 1609 was an important factor in the favoring of a heliocentric system. Galileo's observations indicated that these moons orbited Jupiter and not the Earth or sun directly. If this were the case for Jupiter, Galileo asked, might it not also hold true for the remainder of the solar system? However, the discovery of Jupiter's moons was not in itself sufficient evidence to overthrow geocentrism.

However, there were to be other challenges. Johannes Kepler, a student of Tycho Brahe, had developed an interest in discovering what he believed to be the geometric harmony of the universe. Using his mentor's data, Kepler set to work determining the orbit of Mars (see MARS). Kepler, a firm believer that scientific theories must be supported by facts, did not accept the application of epicycles to the orbits of the planets. To remove the influence of epicycles, Kepler eventually discarded the prevalent ideas of circular planetary orbits in favor of elliptical orbits. This would eventually be incorporated into his three laws of planetary motion (see KEPLER'S THEORY OF PLANETARY MOTION). But more importantly, when Kepler applied his theories to the orbit of the Earth, he found a similar elliptical pattern. This would only be possible in a heliocentric system in which the Earth was in orbit around the sun. With this information Tycho's model was obsolete and by the mid-1600s there were few remaining scientific supporters of geocentrism.

The debate would be finalized with the work of Edmund Halley and Isaac Newton. Prior to Halley, comets were considered to be transitory objects and not in orbit. Halley's investigations of a reoccurring comet indicated that the comet followed an elliptical orbit around the sun (see HALLEY'S COMET). To justify his ideas, Halley required the assistance of Newton's laws of gravitation (see GRAVITATION). Not only did the emerging laws of gravitation support the orbits of comets, it could also serve as the answer to problems with the Copernican model. Gravity was the force that kept objects on the surface of the Earth and the reason why falling objects were attracted to the ground. Indeed, gravity was the governing force for the structure of the universe.

The Ptolemaic theory and its geocentric view of the universe had persisted for over fourteen centuries. While the competing theory of Copernicanism would be established in the 16th century, it would not progress into a workable model until the next century. The 1600s would see a change in scientific thinking from the philosophical basis of the ancient Greeks to a process in which theories must be supported by facts and observations. The contributions of Galileo, Kepler, Newton, and others eventually provided enough evidence to permanently refute geocentrism. In the process they would forever change our view of the universe.

Selected Bibliography

Evans, James. *The History and Practice of Ancient Astronomy.* New York: Oxford University Press, 1998.

Galilei, Galileo. *Dialogue Concerning the Two Chief World Systems.* Translated by Stillman Drake. Berkeley: University of California Press, 1953.

Gingerich, Owen. *The Eye of Heaven: Ptolemy, Copernicus and Kepler.* New York: American Institute of Physics, 1993.

Lloyd, G.E.R. *Early Greek Science: Thales to Aristotle.* New York: W. W. Norton, 1970.

Wilson, Robert. *Astronomy through the Ages: The Story of the Human Attempt to Understand the Universe.* Princeton, NJ: Princeton University Press, 1997.

Hooke's Law (1660–1678): By the mid-17th century the advances in scientific reasoning and experimentation were giving scientists of many disciplines the ability to investigate the natural world. As these researchers applied new tools, such as analytical geometry and calculus (see ANALYTICAL GEOMETRY; CALCULUS), and new instruments, such as the microscope and telescope (see MICROSCOPE; TELESCOPE), it finally became possible to conduct investigations with a significantly higher level of detail and accuracy. While many scientists of this time initially set out to expand on the knowledge of the ancient Greeks, most often that described by Aristotle, they frequently found that their experiments and derived data contradicted many of the existing Greek theories. As the 17th century progressed, scientists became more aggressive in their investigations and ventured into areas of study that previously had not been a focus of science.

One of these areas involved the study of the **elasticity** of an object. An object is said to have elastic characteristics if it will return to its original shape after an applied deforming force has been removed. Studies of elasticity in the 17th century originally started with early investigations on the effects of pressure on a gas. The English physicist Robert Boyle had noted that the volume of a gas is dependent on the pressure applied to it (see BOYLE'S LAW). One of Boyle's employees, Robert Hooke, developed an interest in the ability of air to return to its original volume after the pressure had been removed. Hooke called this the "springiness" of air, and his work in this area became the basis for what would later be called Hooke's Law.

The basic premise of Hooke's Law is that the amount of distortion of an object under stress is directly proportional to the force of the object applying the stress. The amount of distortion is often called the strain on the object and the stress is the outside force being applied. Thus Hooke's Law is frequently paraphrased as "the strain is proportional to the stress." In his research Hooke focused almost exclusively on one-dimensional stresses such as **compression** and stretching. In the natural world stresses are often applied from a number of dimensions, as is illustrated by the stress of bending and twisting. However, Hooke did not limit his study of elasticity to air only. It is known that he studied the elasticity of a number of objects by connecting them to a scale and then attaching one end of the object to a fixed point. Weights, representing the force, were then added to the scale. Using this method it was possible to study the degree of **deformation** in an object in response to a defined force. Hooke conducted experiments on dry wood, long wires, watch springs, and metal helixes. His work with springs deserves a special note, as it is an early example of scientific investigation benefiting industry, as Hooke was able to use his study of elasticity in the design of improved watch springs.

Although there is evidence that Hooke performed these experiments as early as 1660, and even published a brief account of the elasticity of air in *Micrographia* (1665, see CELLS), the majority of his findings were not released until 1678. By this time other investigators were also involved in the study of elasticity. In 1674 the English mathematician William Petty (1623–1687), better known for his study of economic principles, also published some ideas on the nature of elasticity. Petty believed that the elastic nature of an object was due to conformational changes in the object's atomic structure. To Petty, the oscillations of an elastic object after the force had been initially removed were due to imbalances in atomic forces (see ATOMIC THEORY). Once these forces stabilized, the object returned to its original configuration. The renowned English physicist Isaac Newton also made a brief description of elasticity in *Opticks* (1704), although this tended to focus more on elasticity as an attractive force at the molecular level.

Hooke's Law remains an important component of modern studies in physical science. In modern mathematical terms this law can be expressed as the formula:

$$F = kx$$

where F represents the applied force and x is the amount of strain (deformation) measured in the object. The k term is a constant used to define the physical characteristics of the object being tested, for example its composition and length. While in the 17th century Hooke's Law was primarily an indication of a trend to explain natural forces using scientific methods, it still represented a considerable technological advance. Along with studies of magnetism and gravitation it was quickly becoming possible not only to observe the effects of natural forces, but to predict their influence on an object in a test situation. Modern engineers frequently use the principles set forth by Hooke in construction projects or in the design of new materials that can tolerate higher stress levels. From a historical perspective it was Robert Hooke's work on elasticity that provided the foundation for these modern technological advances.

Selected Bibliography

Cajorie, Florian. *A History of Physics.* New York: Dover Publications, 1962.

Centore, F. F. *Robert Hooke's Contributions to Mechanics—a Study in Seventeenth Century Natural Philosophy.* The Hague, Netherlands: Martinus Nijhoff, 1970.

Wolf, A. *A History of Science, Technology and Philosophy in the 16th and 17th Centuries.* New York: Macmillan, 1968.

Human Physiology (ca. 1612–ca. 1681): Physiology is the study of how the cells, tissues, and organs of living organisms function. Prior to the 17th century the majority of the research on the human body was based on anatomy. The ancient Greek scientist Galen established the foundations for this work in the 2nd century and these ideas dominated scientific thought until the time of the Renaissance. In the 16th century, Galen's authority was challenged by the Dutch anatomist and physician Andreas Vesalius. While Vesalius is regarded as one of the prime historical figures in the study of human anatomy, his understanding of human physiology reflects the relative ignorance of the times. He believed that the process of respiration was simply to prevent the blood from overheating, while digestion involved a slow cooking of the food using the heat of the body. Few attempts were made to investigate physiological functions until the time of the 17th century. The limited research that was done was performed primarily by the anatomists and the alchemists. The study of human physiology in the 1600s closely parallels the work being done in the science of chemistry and anatomy. An excellent example of this is the research performed in the 17th century on the structure of the lymphatic system. While anatomists were busy identifying the structure of this circulatory system, little work was being performed to determine its function (see LYMPHATIC SYSTEM). While the 17th century is not often regarded as a time of breakthroughs for physiological science, there were a few important discoveries that illustrated the fact that the sweeping scientific revolution was having an influence. The investigations into human respiration and digestion set the stage for work into the next two centuries. As with many areas of science during the 17th century, the scientists of the times were typically not specialists, but broad-trained interdisciplinary investigators who approached the study of human physiology from a number of perspectives.

One of the leading discoveries on human physiology in the 17th century was the result of the English physician William Harvey's work on blood circulation (see BLOOD CIRCULATION). The circulation of blood had been a focus of physicians since the time of the ancient Greeks. Galen believed that all blood originated in the liver, where it was mixed with "natural spirits," an early concept of nutrition. The blood then passed through veins of the body to the heart where it was mixed with the "vital spirits," or air. This blood was then directly consumed by the tissues and not returned to the liver. While there were a number of challengers to this theory in the late 16th century, by 1628 William Harvey had effectively ended Galen's concept of blood circulation. By using logical analysis, coupled with experimental testing, Harvey demonstrated that blood circulated in a loop between the arteries and veins of the body (see BLOOD CIRCULATION). Later in the century the invention of the microscope (see MICROSCOPE) enabled other researchers, most notably Jan Swammerdam and Marcello Malpighi, to finalize the physiology of the circulatory system with the discovery of red blood cells and capillary beds connecting the arteries and veins. Harvey's discovery of the physiology of the circulatory system marked an end to the ancient Greek reign on human biology. This would have a tremendous impact on the remaining physiological studies of the 17th century.

Early chemical studies of human physiology sought to develop an understanding of the relationship between chemistry and the human body. This early application of chemical principles to medicine is called iatrochemistry. This was popularized in the 16th century by the Swiss alchemist Paracelsus, who believed that many human diseases and problems were due to chemical imbalances within the body. In the 17th century chemistry split into two main factions, the alchemists with their belief in the transformation of matter, and the chemists with their focus on determining the properties of matter (see CHEMISTRY). Many iatrochemists, such as the Dutch physician Franciscus Sylvius (1614–1672), were attempting to derive chemical explanations for physiological processes. Sylvius believed that diseases were due to an imbalance of the acid-base mixture of the blood. Another example is the study of the physiological process of human metabolism. Metabolic processes are frequently studied by both biologists and chemists and this represents a field in modern science called **biochemistry**. One of the earliest researchers of human metabolism was the Italian scientist Sanctorius (1561–1636). To study metabolism Sanctorius designed a special "weighing chair." In an experiment lasting almost three decades, Sanctorius measured the weight of all the food and liquids he consumed as well as the waste material he excreted. While this experiment was far from scientifically sound, Sanctorius did note that the weight of his excrement did not equal the weight of the ingested material. He concluded that the remaining material was lost by a process called "insensible perspiration," although he did not identify its nature. However, Sanctorius is credited with developing two instruments that were used throughout the remainder of the century for the study of human physiology. These were the medical thermometer (see THERMOMETER) and a pendulum device used for more accurately measuring pulse rate.

Another area of physiological research that was conducted in the 17th century was on the process of human digestion. Historically the digestive system has been credited with a wide array of functions other than the processing of the food. Galen believed the stomach to be a site where the blood was cooked prior to being delivered to the tissues. The 17th-century iatrochemist Jan Baptista van Helmont believed that an organ of the digestive system that he called the *archeus* regulated all of the physiological processes of the body. Unfortunately, such an organ does not exist. In modern medicine this function is assumed by a number of glands, including the liver, hypothalamus, and thyroid gland. However, Helmont's contributions were not all negative. He recognized that the acid in the stomach was involved in the digestive process. He also associated the gallbladder with digestion, although he considered its role to be that of neutralizing acids. Furthermore, as an iatrochemist he considered the stomach to be the site of a fermentation process, a contradiction to the still-supported Galenic theories. The iatrochemist Franciscus Sylvius contended that the secretions of the salivary glands, pancreas, and gallbladder were involved in digestion, not just the acid of the stomach. The work of Helmont and Sylvius would be expanded greatly in the 18th century with an increased reliance on the gathering of experimental evidence.

Experimental research into the process of human respiration was also conducted during this time. The study of the physiology of respiration was complex in that scientists were just beginning to understand the process of blood circulation (see BLOOD CIRCULATION) as well as the nature of gases and combustion (see COMBUSTION; GASES). It was well known that air was required for life, and that the process of respiration moved air in and out of the body. Robert Boyle experimentally proved this around 1660 using experiments with mice and flames (see GASES). While some still believed that the function of respiration was to cool the blood, most scientists of the 17th century supported the idea that respiration was involved with the process of combustion within the body and that the respiration process may serve to remove some of the waste products of this combustion. The English physician Richard Lower conducted one of the more important experiments in respiration physiology during this time. Lower was interested in determining the nature of the color difference between arterial and venous blood. He duplicated a series of experiments originally performed by Vesalius and Robert Hooke in which the **thorax** of a dog was cut open and a bellows attached to control respiration. In his own experiments Lower noticed that when blood from the vena cava, the main vein from the body to the heart, was forced through the lungs, and air was passed over it by the bellows, the color of the blood returning to the heart via the pulmonary vein was bright red. Lower then proposed that this air, or "nitrous spirit" as it was called then (see COMBUSTION; GASES), leaked from the blood into the tissues of the body. When this happened, the blood lost its red color. The fact that this gas was oxygen would not be conclusively proven until the work of Antoine Lavoiser in the 18th century. John Mayow conducted a number of additional experiments to identify the nature of the nitrous spirit (see COMBUSTION).

Many scientists of the 17th century subscribed to the belief that the human body was simply a sum of its parts. This allowed the functions of the body to be investigated and described as one would a machine. This theory was called the mechanist theory and supporters of this idea made a number of contributions to the study of human physiology in the 17th century. Many of the scientists previously mentioned held mechanist viewpoints, but perhaps the two greatest mechanists to make contributions were the French philosopher René Descartes and the Italian physician Giovanni Borelli (1608–1679). Descartes was the leading mechanist of the time and his most significant contribution is considered to be distinguishing between voluntary and involuntary reactions of the body. To Descartes, the **pineal gland** was the organ that commanded the actions of the body. The pineal gland was the source of the nervous fluid and it was this fluid that directed voluntary responses such as muscle contraction to occur. Other functions, such as the **peristaltic** contractions of the intestines, were automated responses. Descartes considered the body to be a self-regulating machine with the majority of the functions under automation. Descartes then extended this mechanist theory of biology to more philosophical discussions on the nature of life and the human soul. For this he is frequently regarded as one of the greatest philosophers in history.

In 1681, Giovanni Borelli published *De Motu Animalium*, which outlined his mechanist views on the physiology of motion. While this book presented theories of motion in a wide variety of animals, it primarily concerned itself with the nature of human motions such as running and walking. Borelli viewed muscles as levers, and thus may be described using physical laws. Previously, the Dutch anatomist Nicolaus Steno had conducted some detailed microscopic examinations of muscle contraction and noted that it was the action of individual muscle fibers that was responsible for muscle action. Borelli extended this to include a chemical factor. He suggested that the nervous system provided a "juice," or chemical, to the muscle fibers. This in turn caused a chemical reaction within the fibers that resulted in a shortening of the fibers. He applied this theory to some of the earliest studies on the contraction of the heart.

The process of vision attracted a significant amount of attention during the 17th century. The majority of this work was performed by physicists trying to understand the principles of optics, specifically the refractive and reflective properties of a lens (see OPTICS). Many of these researchers simply related the human eye to a type of primitive lens. However, the German physicist Johannes Kepler recognized that the physiology of the eye was more complicated than a single-lens system. Kepler predicted that the purpose of the lens was to focus light from different points onto the retina, which in turn recorded the image. Kepler noted that after exposure to a bright light source, an artifact of the image frequently may be seen for some time. He further proposed that the eye sends signals to the brain, which then interprets the image. These ideas were far ahead of their time and closely parallel modern theories of vision.

It should be noted that physiological studies were not confined to humans in the 17th century. A number of researchers, most notably the Italian scientist Marcello Malpighi, were conducting studies on animal physiology. The majority of this work was conducted on insects (see ENTOMOLOGY) and dogs. Many areas of human physiology research were aided by physiological studies in animals. A unified attempt to understand human physiology would first require some substantial improvements in the field of chemistry. In the late 18th and early 19th centuries the advances in medicine, chemistry, and general scientific procedures would result in the expansion of our understanding of human physiology. However, many of the principles of respiration, digestion, the action of muscles, and the nervous system necessary for these later advances were pioneered by 17th-century physiologists.

Selected Bibliography

Duffin, Jacalyn. *History of Medicine: A Scandalously Short Introduction.* Toronto: University of Toronto Press, 1999.

Serafini, Anthony. *The Epic History of Biology.* New York: Plenum Press, 1993.

Taton, Rene, ed. *The Beginnings of Modern Science.* New York: Basic Books, 1958.

Wolf, A. *A History of Science, Technology and Philosophy in the 16th and 17th Centuries.* New York: Macmillan, 1968.

Hydrologic Cycle (1674–ca. 1693): Hydrology is the study of how water is distributed and circulates on our planet. The importance of water to all living organisms was well known by the scientists of ancient civilizations as early cultures were dependent on the availability of water for all aspects of their lives. For this reason early scientists and philosophers were often tasked with explaining the movement of water between the environment and all living things. In the modern scientific community the natural movement of water, called the hydrologic cycle, is known to include the action of the atmosphere and oceans as well as streams, rivers, and springs. However, until the 17th century the interaction of these natural phenomena was not recognized and often poorly interpreted. Almost every ancient culture possessed its own version of the hydrologic cycle. In the Western world the ancient Greek philosopher and mathematician Thales of Miletus proposed that the surface of the Earth actually floated on an ocean of water. Furthermore, he considered water to be one of the basic elements of all matter. This was expanded on by later generations of Greek scholars. In the science of Aristotle, water was one of the four primal elements along with fire, air, and earth (see CHEMISTRY). In his work on the hydrologic cycle Aristotle was responsible for the synthesis of many previous ideas. He believed that the originating point of all rivers and streams was a vast subterranean sea. As the concept of water vapor in the atmosphere was unknown during this time, the level of this sea was replenished by the condensation of elemental air to water. While Aristotle did recognize that precipitation had some influence on the level of rivers and streams, he did not consider it to be a major factor in the global movement of water. It was Aristotle's views that dominated hydrologic thinking until the time of the 17th century.

While the period of time between Aristotle and the Renaissance is not often recognized for its scientific advances, science was still being conducted, although at a much-reduced level. This was also the case for the study of hydrology. However, most of the work done during this time made relatively minor contributions or simply served to restate existing Aristotelian principles. The development of quantitative science during the late 16th and early 17th centuries resulted in a redirection of thinking in the science of hydrology. While there were a significant number of people who made contributions during this time, the majority of the effort focused on the work of three investigators, Pierre Perrault, Edmé Mariotte, and Edmund Halley.

Pierre Perrault was a French hydrologist who studied the movement of water in the Seine River area located just outside the French city of Dijon. In 1674 Perrault published *The Origin of Springs*, which summarized his studies of the region. Perrault was convinced that Aristotle's theory that streams and rivers were regenerated from underground sources was flawed, as it did not take into consideration the influence of precipitation. Aristotle thought that rainfall made a minor contribution to the regeneration of the underground sea, the primary source for rivers and streams. Perrault proposed that the rainfall itself was sufficient to explain the flow of these waterways. He chose a section of the Seine River for an experimental verification of this idea. First, Perrault identified the area of the watershed for his experimental area of the river. Basically, a watershed is all of the input sources (runoff, springs, streams, etc.) that contribute to the flow of a larger body of water. He then determined the amount of rainfall for the area, about 19.3 inches. When this amount was multiplied by the area of the watershed, Perrault was able to determine the total amount of water available. He then estimated the flow rate of the river in a given period of time. When he compared the inputs to the watershed versus the water flow out of his study area, he noted that only about one-sixth of the total rainfall was actually entering the river. While Perrault did not take into consideration evaporation or infiltration of the water into the lower rock layers, he did demonstrate that the amount of rainfall alone was enough to support the flow of water in this river. Using the information that he had obtained experimentally and mathematically, Perrault proposed that all rivers and streams were sustained by the action of precipitation. This represented a major change in thinking for the study of hydrology.

A second French scientist, Edmé Mariotte, made a number of significant contributions during the mid-17th century to the study of hydrology and hydraulics, or the relationship between pressure and the physical properties of a liquid (see HYDROMECHANICS). On the origin of springs, Mariotte expanded upon Perrault's work. Perrault could not prove that there was an adequate amount of rainfall infiltrating the ground to recharge groundwater sources. However, Mariotte demonstrated that the flow from springs was directly related to the amount of precipitation. During dry times the flow from springs decreased, with the opposite occurring during the time of plentiful rainfall. Mariotte suggested that precipitation falling on a hill or mountain percolated down into the ground until it reached an impermeable layer. It then moved along this layer until it found a weak location in the rock where it could return to the surface. Mariotte also developed

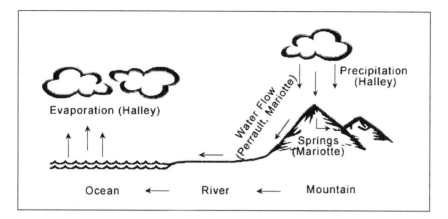

Figure 10. The hydrologic cycle noting the contributions of Halley, Perrault, and Mariotte. The complete path was first proposed by Edmund Halley.

a more accurate method of determining the discharge of a stream or river. By using wax floats he was able to calculate the velocity of water in a channel. In doing so he made the additional discovery that the velocities of water at the top and bottom of the channel are not necessarily the same. He thought that it was this difference that explained some of the dynamics behind stream structure. He then utilized the ability to estimate flow velocity to once again measure the impact of precipitation on stream flow. Over a three-year period he measured the precipitation and flow rates of a watershed of the Seine River near Paris. His conclusions mathematically confirmed the work of Perrault that there was a sufficient input of water from the watershed to explain the flow of the river. This effectively ended any remaining support for the theory of a subterranean sea.

Edmund Halley was an English astronomer and mathematician who is best recognized for his study of comets (see HALLEY'S COMET). However, Halley made a number of other important contributions to the advancement of science during the 17th century, including hydrology. Halley's interest in hydrology focused on the process of evaporation. It was Halley who completed the hydrologic cycle begun by Perrault and Mariotte (see Figure 10). Halley explained that the heat of the sun caused molecules of water to expand and become lighter than air; thus, the warmer the air, the more water it may contain. As the air cooled, the reverse process occurred and the water would leave the air as dew or rain. Halley suggested that the heating of the oceans created vast clouds of vapor that drifted inland by the action of wind.

Precipitation results when this vapor encounters the cooler temperatures associated with a mountain. This precipitation then infiltrates into the mountain and becomes a spring in much the same manner as described by Mariotte. Halley then set out to prove that the amount of evaporation was sufficient to generate the precipitation needed to maintain the flows of rivers and streams. As an experimental system he chose the Mediterranean Sea and the nine major rivers that flow into it. He estimated that the rivers contribute almost 1.827×10^9 tons of water per day to the Mediterranean while evaporation accounted for the loss of 5.28×10^9 tons on a summer day. Thus, the amount of water leaving the ocean was more than sufficient to provide for the flow of the rivers. Finally, Halley noted that for inland seas that did not have an exit point to larger oceans, for example the Dead Sea, the process of evaporation should be making these seas continuously more salty as the waterways carried minerals to the sea and evaporation removed the water. He even projected that this observation could be used to estimate the age of the Earth.

Once again in the 17th century the application of the scientific method, where ideas are supported by experimental evidence, had served to overturn a theory that had persisted previously for centuries. The study of hydrology entered the century as a science based upon conjecture and philosophical discussions. By applying basic mathematical principles, supported by logical experimentation, Perrault, Mariotte, and Halley made significant advances on the study of hydrologic principles. And while some of their conclusions were not entirely correct, for example Halley's view on the molecular basis of evaporation, they did represent important scientific advancements of the 17th century.

Selected Bibliography

Biswas, Asit K. *History of Hydrology*. Amsterdam, Netherlands: North-Holland Publishing, 1970.

Gohau, Gabriel. *A History of Geology*. New Brunswick, NJ: Rutgers University Press, 1990.

Taton, Rene, ed. *The Beginnings of Modern Science*. New York: Basic Books, 1958.

Hydromechanics (ca. 1644–1660): The 17th century was a time when scientists of many disciplines were attempting to explain the physical laws of nature. This required a revolution in the scientific world as researchers incorporated new methods of scientific reasoning and experimentation into their work. This change was made possible by new mathematical procedures (see ANALYTICAL GEOMETRY; CALCULUS) and advances in the invention of scientific instrumentation (for example, see MICROSCOPE; PENDULUM CLOCK; TELESCOPE). As with many of the scientific advances of the scientific revolution, the roots of the 17th-century work in hydromechanics began during the time of the ancient Greeks. Hydromechanics is an area of science that examines the effects of forces, typically pressure, on a fluid. The mechanics of fluids was not the only area being examined during this time. Similar studies were being conducted on the effects of pressure on air (see BOYLE'S LAW). Hydromechanics may be divided into two areas, hydrostatics and hydrodynamics. Hydrostatics examines the effects of pressure on a fluid at rest, while **hydrodynamics** focuses on the physical properties of a fluid in motion. The study of hydraulics is the application of hydrodynamic principles. In the 17th century the majority of the research was conducted on hydrostatics, although there were a few contributions in the area of hydrodynamics that were expanded on by 18th-century scientists.

The study of hydromechanics can be said to have begun with the work of the Greek mathematician Archimedes in the 3rd century B.C.E. Legend has it that Archimedes was tasked with determining whether the king's crown was made of pure gold or contained silver impurities. The methodology must be such so that the crown was not harmed in the process. While sitting in his bath, Archimedes realized that an object displaces water and that the amount of displacement is a function of the density of the object. Since gold and silver have different densities, it was possible to calculate what the displacement of the crown should be if it were made of gold by first comparing it to an object of pure gold with a weight comparable to that of the crown. By comparing the displacement of the crown versus this object, it was possible to determine the purity of the crown. Archimedes refined this into what is now called Archimedes' Principle. This law defines buoyancy as the force of the fluid that opposes gravity. This force is equal to the weight of the fluid displaced by the floating body, which in turn equals the weight of the object in the water. In other words, the buoyant force is equal to the weight of the object. This explains why boats may have cargo added to them without sinking. As the weight of the boat increases, so does the buoyant force. This finding would have a major influence on 17th-century hydromechanics.

The study of hydraulics in the 17th century is directly connected with the invention of the barometer by the Italian scientist Evangelista Torricelli. Torricelli, a student of Galileo's, inherited his mentor's interest in fluid mechanics. Galileo had developed an interest in the design of water pumps to remove water from the bottom of mine shafts (see BAROMETER; VACUUM PUMP). In doing so he noted that there was a limit to the height of a column of water (thirty feet) and suggested that there may be a relationship between the density of the liquid and the height of the column. However, it was Torricelli who conducted the experiments. In 1644, together with Vincenzo Viviani, he invented the "Torricelli Tube." This was an instrument that was initially designed to study vacuums, but actually it served to measure the effects of air pressure (see BAROMETER). In a barometer the downward pressure of the atmosphere forces a liquid to a height in a tube dependent on the density of the liquid and the weight of the atmosphere. Since the density should be a constant, any variations in the weight of the atmosphere may be detected.

The French physicist Blaise Pascal made a number of improvements to the barometer and was the first to employ the instrument as a mechanism for studying the effects of pressure on a liquid. Pascal made a number of contributions to 17th-century science and mathematics (for example, see BAROMETER; PROBABILITY), but as a physicist what he is most remembered for are his discoveries in the field of hydrostatics and the derivation of the principle that bears his name. Pascal's Principle, which is sometimes called Pascal's Law, states that for a liquid at rest, the pressure at any point in the fluid is the same in all directions. Furthermore, any change in the pressure at that point is conducted without loss to all other points within the fluid and along the walls of the container. Pascal's Principle was actually first developed by the Dutch inventor Simon Stevin, or Stevinus (1548–1620), at the end of the 16th century. Stevinus had investigated what is called the "hydrostatic paradox." Stevinus had noted that the force of a fluid on the bottom of a container was dependent on the size of the container's bottom and the height of the fluid above this surface. Pascal refined the mechanics of this idea and applied the hydrostatic principles to the movement of a fluid under pressure.

Pascal noted that the amount of pressure is a function of the strength of the applied force divided by the size of the area that the force is being directed upon. Using this information, Pascal proposed the development of a hydraulic press. In a hydraulic system (see Figure 11), the operator applies force to a piston containing a liquid (or air). Since the pressure in a liquid is the same at every point within the

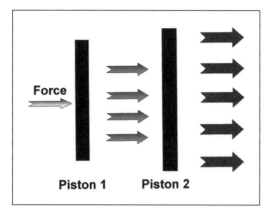

Figure 11. The principles of a hydraulic press. The initial force applied to piston 1 is felt evenly along its entire surface. This force is amplified and applied through the medium to the surface of piston 2. Since every point along the surface of piston 2 feels the same force, the result is a significant amplification of the initial force.

liquid, the strength of the force on the wall of the piston is dependent on the surface area of the piston. The pressure applied by the operator is then transferred to a second piston. Although the applied pressure is the same, if the second piston possesses a larger surface area, there will be more particles exerting this force on the face of the piston, and the strength of the force will be increased. If the second piston has a surface area that is four times larger, the force will be four times greater. This magnification of force serves as the principle behind the operation of hydraulic machinery.

Pascal's contributions to hydromechanics also influenced studies on air pressure (see BOYLE'S LAW). By recognizing that pressure affects the level of a fluid, Pascal was able to use a barometer to measure air pressure at various altitudes (see BAROMETER). He observed that there was a direct relationship between altitude and atmospheric pressure that allowed him to make some early predictions on the height of the atmosphere. This type of information allowed scientists to construct tables that indicated the altitude above sea level for specific barometric pressures. For example, a century after Pascal, Thomas Jefferson (1743–1826) used the observed differences in barometric readings he made to calculate the altitudes of the area surrounding his home.

The study of hydrostatics represented an important advance for the science of the 17th century. As with air pressure, the physical laws of nature were being observed and quantified using the experimental method. Once hydrostatic principles were understood, it was possible to develop a more detailed explanation of the nature of fluids in motion, or hydrodynamics. In the 18th century Daniel Bernoulli conducted an extensive investigation of hydrodynamic principles. *Hydrodynamica,* published in 1738, presented what is called the Bernoulli Principle. This states that the pressure of a fluid in motion decreases as the velocity decreases. He also performed some of the earliest experiments on the effects of temperature on the motion of a liquid. He noted that an increase in temperature resulted in a subsequent increase in pressure and vice versa. These discoveries would have a strong influence on investigations of fluid mechanics well into the 20th century.

Selected Bibliography

Taton, Rene, ed. *The Beginnings of Modern Science.* New York: Basic Books, 1958.
Wolf, A. *A History of Science, Technology and Philosophy in the 16th and 17th Centuries.* New York: Macmillan, 1968.

Hygrometers and Hygroscopes (ca. 1657–ca. 1685): The relative humidity of the atmosphere is the amount of moisture in the air in comparison to the maximum or saturated amount that it can hold. Since the amount of moisture is directly associated with the stability of the atmosphere, and thus the development of weather patterns, calculations of humidity have been an important component of meteorological science since its inception in the 17th century (see METEOROLOGY). The instrument used to measure the humidity of the atmosphere is the hygrometer, derived from the Greek words for moisture and measure. While in modern meteorology there are several different types of hygrometers, from those that use electrical methods of determining humidity to those that measure dew points, each of these instruments can trace its history to the 17th century. During this time a number of scientists experimented with the invention of instruments to accurately measure humidity.

A simple hygrometer may be constructed by placing a dry sponge outside on a balance and recording the change in weight over a predetermined amount of time. Such devices were proposed as far back as the 15th century by a number of scientists, including the Italian inventor Leonardo da Vinci. However, in the 17th century there developed a strong interest in the invention of more precise scientific instruments to support the quantitative measurements being performed in most scientific disciplines. The study of meteorology was no exception. The development of the hygrometer paralleled the invention of instruments to measure temperature and barometric pressure (see BAROMETER; THERMOMETER). Sometime around the middle of the

17th century (ca. 1657) members of the Academia del Cimento, an Italian scientific society located in Florence, designed a simple hygrometer that used the process of condensation to measure humidity. This device consisted of a cork funnel that was lined on the outside with tin. When the cork cone was filled with ice, the water in the atmosphere condensed on the tin. The individual drops of water were counted in early models. Later, the quantity of water was determined by collecting it in a graduated measuring device. As with many of the scientific instruments of the 17th century, there existed a considerable amount of variation in the design of hygrometers. Many of these devices functioned on the premise that as objects absorbed water from the atmosphere they would increase in size. Among the inventors of this style of hygrometer was the French physician Guillaume Amontons (1663–1705), better known for his work with the thermometer (see THERMOMETER). This instrument measured the changes in size of a small wood sphere in relation to variations in atmospheric humidity. Others suggested measuring changes in the weight of chemicals, such as sulfuric acid, in response to changes in humidity.

As the 17th century progressed, a number of scientists were in the process of developing hygrometers that mechanically measured changes in humidity. These devices are commonly called hygroscopes, although this term is frequently used interchangeably with hygrometer. Robert Hooke is often credited with the invention of the first mechanical hygroscope. Described in his 1665 publication *Micrographia* (see CELLS), Hooke's hygroscope actually utilized many of the same principles as the previous devices. This hygroscope used a portion of a wild oat plant as a detecting device for the instrument. It was widely known that as an oat husk is exposed to moisture certain portions of the husk undergo a change in their orientation in relation to the remainder of the plant. Hooke proposed that this change be measured by anchoring one end of the husk to a piece of wood and attaching a pointer, actually a small pin, to the other end. The pin was oriented on another piece of wood that had etched upon it a dial to measure movement of the pin. As the humidity changed, the resulting changes in the oak husk would be transferred to the pin and measured on the dial. An unknown Irish inventor proposed a similar instrument at about the same time except that this instrument measured the changes in relative humidity by the shrinkage or expansion of two wood boards. Once again, the changes in the size of the wood were transferred mechanically to a pointer that indicated the change on a preestablished scale.

The hygrometers and hygroscopes of the 17th century were obviously primitive in design and as such contained some significant problems. Comparison of measurements between the different devices must have been next to impossible. Furthermore, as the expansion and contraction of the organic material being used was dependent upon the characteristics of the individual plant from which it was derived, calibration of the instruments must also have presented a special challenge. Improvements to the hygroscope would be made over the next few centuries, including some important work on the precision and accuracy of the hygroscope in the late 18th century. For example, around 1770 Horace Benedict de Saussure (1740–1799), a Swiss physicist, began using a human hair in hygrometer design. The human hair proved to be very susceptible to changes in humidity and greatly increased the sensitivity of the instrument. Despite the inadequacies of the 17th-century instruments, the invention of the hygrometer does mark an important turning point in the study of weather. When coupled with the invention of thermometers, barometers, and wind gauges, scientists were now able to make quantitative measurements of weather-related phenomena (see METEOROLOGY). This result was the birth of meteorology as a scientific discipline.

Selected Bibliography

Bud, Robert, and Deborah Jean Warner, eds. *Instruments of Science.* New York: Garland Publishing, 1998.

Wolf, A. *A History of Science, Technology and Philosophy in the 16th and 17th Centuries.* New York: Macmillan, 1968.

J

Jupiter (1610–1687): Jupiter, the fifth and largest planet of our solar system, has dominated the night sky throughout recorded history. Typically the second brightest planet when viewed from Earth, Venus being the first, Jupiter has held a key position in mythological lore since the time of the Roman Empire. To the Romans, Jupiter was the son of Saturn and the god of the sky and weather. Legend has it that after defeating his father, Jupiter became the king of the gods and held dominion over the fate of humans. In Greek mythology Jupiter was represented as the god Zeus. To the ancient Chinese civilizations Jupiter represented one of the five elements. The 11.86-year orbit of Jupiter also closely matched the twelve lunar cycles in a year and thus held special importance in the Chinese belief of natural cycles. While the planet itself was not discovered in the 17th century, Jupiter would play an important role in the verification of the new theories on the physical laws of the universe that were developing during this time.

Perhaps Jupiter's greatest contribution to the science of the 17th century would begin centuries earlier with the Greek astronomer Ptolemy's ideas on the structure of the universe. Ptolemy proposed that the universe was centered on the Earth and that in this geocentric model all other objects orbited the stationary Earth. While the stars, sun, and moon cooperated with this model, the planets presented real challenges to the theory. Jupiter's path consisted of periods when the planet appeared to be stationary against the backdrop of the night sky as well as periods when it displayed a retrograde, or backward, motion. While other planets displayed more pronounced retrograde motions, especially in the case of Mars (see MARS), Jupiter's motions would severely challenge the Ptolemaic system. In order to make the observations match the models, Ptolemy compensated by adding epicycles, or small circular motions, to the orbit of Jupiter (see KEPLER'S THEORY OF PLANETARY MOTION). These epicycles greatly complicated the Ptolemaic system and would encourage the development of theories with fewer sources of error. Chief among these was a heliocentric, or sun-centered, model made popular by Nicholas Copernicus in the 16th century (see HELIOCENTRISM). The simplicity of this model, in which the Earth and all other objects were in orbit around the sun, created considerable controversy in the 17th century.

One of the primary supporters of Copernicus's ideas was the Italian astronomer Galileo. The invention of the telescope in 1609 revolutionized observational astronomy (see TELESCOPE) as it began to reveal the complexities of the universe. When Galileo turned his primitive telescope on Jupiter, he noticed that there were four smaller objects surrounding the planet. Originally Galileo thought them to be stars, but after more careful observations he determined that these objects were actually in orbit around the planet. Collectively these four moons would be called the Galilean moons in honor of their discoverer. Later, the German astronomer Simon Marius (1573–1624) named the moons Io, Europa, Ganymede, and Callisto, all associates of the Greek god Zeus. More importantly, however, the discovery of moons other than the one around Earth would deal a severe blow to geocentricism. The primary support for the Ptolemaic system rested on the fact that the moon was in orbit around the Earth, Ptolemy simply extended this premise to include all heavenly bodies. The discovery of Jupiter's moons provided a verification of the heliocentric theory. If one substituted the sun for Jupiter and the known planets for the four moons, then it became possible to visualize a system centered on the sun (see HELIOCENTRISM). In 1610 Galileo published his studies of Jupiter in *The Messenger of the Stars*. Galileo was a relentless opponent of any scientific reasoning that was not supported by experimentation or observation. To Galileo, it had become unacceptable to accept a geocentric model for which there was little scientific support. In the two decades following his discovery, Galileo launched repeated attacks on the defenders of Ptolemaic thought. In 1632 Galileo published *Dialogue on the Two Chief World Systems—Ptolemaic and Copernican*. This book would create considerable controversy with the ruling Roman Catholic Church, whose doctrines favored a universe centered on the Earth and man. Under considerable pressure, in the last years of his life Galileo was forced to recant his statements. However, his efforts had not gone unnoticed by the scientific community. In the span of a few decades, the Galilean moons would undermine a belief system that had persisted for over fourteen centuries. In its place would be a model that was supported by observation and experimentation.

A number of other important observations were made of Jupiter in the 17th century. Improvements in the telescope as the century progressed made it possible to view Jupiter in greater detail. In 1664 Robert Hooke, an English scientist, would be the first to describe a large spot, later to be called the Great Red Spot. In 1672 the French astronomer Giovanni Cassini's observations of Jupiter confirmed Hooke's findings. While the Great Red Spot has varied in both size and intensity since it was first noticed, it currently covers an area of approximately 141 million square miles. With these more powerful telescopes it was now also possible to detect colored bands around the planet. By examining patterns in the bands, Cassini was able to calculate that it took 9 hours and 56 minutes for Jupiter to revolve on its axis, fairly close to the modern value of 9.85 hours. A similar strategy was used by Cassini to calculate the rotation of Mars (see MARS).

Studies of Jupiter in the 17th century would not be confined to observational astronomy only. In the later half of the century Jupiter became a testing ground for many of the developing theories on the physical laws of the universe. Early in the century Johannes Kepler had observed that the recorded information on the orbit of Mars did not support a circular orbit. Instead, when the orbits of Mars and Earth were compared, the orbital pattern appeared to be elliptical in shape. His work eventually led to the establishment of the Keplerian laws of planetary motion (see KEPLER'S THEORY OF PLANETARY MOTION). These laws would be the dominant theory in astronomy until the time of Isaac Newton. While Kepler had correctly established the shape of planetary orbits, he mistakenly thought that the force that kept the planets contained in these orbits was magnetism (see MAGNETISM). Later in the century the Italian astronomer Giovanni Borelli used the moons of Jupiter to once again investigate the nature of this force. By 1666, Borelli had made sufficient observations to expand on Kepler's ideas. Borelli contended that the force that established the orbits of the planets consisted of three primary components. The first force was derived from the natural movement of the planets away from the sun by centrifugal force. The second influential force was directed outward from the sun along the plane of the sun's rays. The final force was the natural attractive force of the sun. Borelli summarized that the observed orbits of the planets, and the moons of Jupiter, were a result of these forces being in equilibrium. While Borelli was advanced in the description of the nature of these forces, it would not be until the time of Isaac Newton that the underlying factor, gravitation, would be described (see GRAVITATION).

In his observations of Jupiter, Cassini had noted that the revolutions of the moons of Jupiter occurred at a specific rate. While others, such as Galileo, had noted that it may be possible to measure time by determining when the orbit of each moon was eclipsed by the planet, it was Cassini who would first attempt to formulate tables for the events. However, he noted that for each moon the period of revolutions varied as the distance between the Earth and Jupiter changed. The Danish astronomer Olaus Roemer (1644–1710) suggested that this observation was due to the fact that it took light longer to reach us when the Earth and Jupiter were on opposite sides of the sun in comparison to a shorter period of time when both were on the same side of the sun. While there had been attempts earlier in the century to measure the speed of light between two points on the Earth (see LIGHT SPEED), efforts to use the larger distances of space had been hampered by inaccurate distance and time measurements. However, by 1675 these problems had been resolved by the invention of the micrometer (see MICROMETER) and pendulum clock (see PENDULUM CLOCK). A few years earlier, Cassini had used these instruments to calculate the distance of the Earth from the sun, and thus the width of the Earth's orbit. Knowing that the difference in the observed eclipses of Jupiter's moons was twenty-two minutes, and recognizing that the difference in the distance between these two observations was the width of the Earth's orbit, Roemer calculated the speed of light to be about 140,000 miles per hour (see LIGHT SPEED). While only about 75% accurate, it was an important first calculation made possible by advancements in instrumentation.

As the century drew to a close, Isaac Newton indirectly used Jupiter as a verification of his theories on gravitation. During his observations of Jupiter, Cassini had noted that the banding patterns along the equator took a longer time to complete a rotation than did the poles of the planet. Thus, the planet was widest at the equator, a phenomenon called an equatorial bulge. In 1670 Jean Picard (1620–1682), Jean Richer (1630–1696), and Cassini had noted that time measurements using a pendulum clock varied when readings were taken near the Earth's equator. Since the pendulum clock works by gravitation (see PENDULUM CLOCK), it was believed that measurements at any point on the Earth's surface should be constant. However, Newton demonstrated that as the distance from the Earth increases, the force of grav-

Table 2
Satellites of Jupiter in order of discovery. Location indicates the order of the moon from the planet.

Name	Year Discovered	Location	Discoverer	Nationality
Callisto	1610	8th	Galileo	Italian
Europa	1610	6th	Galileo	Italian
Ganymede	1610	7th	Galileo	Italian
Io	1610	5th	Galileo	Italian
Amalthea	1892	3rd	Edward Bernard	American
Himalia	1904	10th	Charles Perrine	American
Elara	1905	12th	Charles Perrine	American
Pasiphae	1908	15th	Philibert Melotte	French
Sinope	1914	16th	Seth B. Nicholson	American
Carme	1938	14th	Seth B. Nicholson	American
Lysithea	1938	11th	Seth B. Nicholson	American
Ananke	1951	13th	Seth B. Nicholson	American
Leda	1974	9th	Charles Kowal	American
Adrastea	1979	2nd	Voyager 2 probe	American
Thebe	1979	4th	Voyager 2 probe	American
Metis	1980	1st	Voyager 2 probe	American

ity decreases. As a planet rotates, centrifugal force causes a bulging at the equator similar to what Cassini had observed. Thus, the gravitational force at points along this bulge should be less, which would be noted by variations in the timing of the pendulum clock (see GRAVITATION).

Since the time of Galileo twelve additional satellites of Jupiter have been detected, although the next discovery would take almost three centuries and some of the smaller satellites would require the use of spacecraft to detect (see Table 2). In the 20th century humans sent a number of unmanned space probes to the planet in an attempt to learn its secrets. The Voyager 1 and 2 probes detected new moons and the presence of a planetary ring similar to the one around Saturn, but less pronounced (see SATURN). The Galileo space probe went into orbit around Jupiter in 1995 and since that time has collected tremendous amounts of data on the composition of the gas giant's atmosphere and the nature of the Galilean moons. More recently, the Cassini space probe will conduct examinations of Jupiter in 2001 while on its voyage to explore the moons of Saturn (see SATURN).

Modern telescopes are immensely more powerful than any instruments of Galileo's time. Today's astronomers have the ability to monitor objects from the perspective of any point along the electro-magnetic spectrum (for example, visible light, ultraviolet radiation, microwave, and X-ray). And while these instruments tend to be focused on distant stars, black holes, and quasars, there have recently been a number of events that have focused attention in our own backyard. As it was in the 17th century, Jupiter has once again dominated the night sky. In 1994, Comet Shoemaker-Levy made a spectacular collision with Jupiter. Witnessed by a host of telescopes and space probes, it may be the first recorded observation of a cometary impact, although there is evidence that Cassini may have witnessed a similar event on Jupiter in the 17th century. Coupled with this is the recent discovery of ice on the moon Europa, and mounting evidence that this moon may harbor a liquid ocean beneath the ice. Studies of Jupiter and its moons may provide indications on the early evolution of our solar system and life on Earth. It seems appropriate that the planet that served as a testing ground for the scientific revolution of the 17th century is now the lab-oratory for a new generation of scientists exploring the fundamental principles of life and evolution.

Selected Bibliography

Galilei, Galileo. *Dialogue Concerning the Two Chief World Systems.* Translated by Still-man Drake. Berkeley: University of California Press, 1953.

Motz, Lloyd, and Jefferson H. Weaver. *The Story of Astronomy.* New York: Plenum Press, 1995.

Spangenburg, R., and D. K. Moser. *The History of Science from the Ancient Greeks to the Scientific Revolution.* New York: Facts on File, 1993.

Wilson, Robert. *Astronomy through the Ages: The Story of the Human Attempt to Under-stand the Universe.* Princeton, NJ: Princeton University Press, 1997.

K

Kepler's Theory of Planetary Motion (ca. 1600–1687): Building on the astronomical observational foundations of the Babylonians, and inspired by their new quest to explain the forces of nature, the ancient Greeks were among the first recorded civilizations that attempted to mathematically explain the motions of the sun, moon, and planets. The motions of the sun and moon followed calculable patterns, and early calendars exploited the predictable nature of these heavenly objects for many uses, primarily the synchronizing of agriculture with the seasons. The planets, however, were another matter. The term "planet" is derived from the Greek word *planaomai*, a verb meaning to wander. And the planets were indeed wanderers of the night sky. Unlike the sun, moon, and stars, the planets appeared to be able to change not only their speed of movement but also their direction of motion in the night sky. Of special concern was Mars, whose movement consisted of stationary phases, called stations, and periods in which the planet moved backward across the sky, called retrograde motion. To the ancient Greeks, the objects of the night sky were of perfect origin, and thus the erratic motion of the planets presented a special problem.

The Milesian philosophers of ancient Greece, a group that included Thales and Anaximander, were among the first to construct a mechanical model for the heavens. This model consisted of three rings that housed the sun, moon, and stars. These rings surrounded the Earth, which itself was portrayed as a large cylinder with a flat surface on top. The rings were spaced at distances nine, eighteen, and twenty-seven times the diameter of the Earth. The Milesians, however, made no provisions for planets in their model. It was not until the time of

Plato that an effort would be made to distinguish between the movement of the stars, moon, and sun and the more random movement of the planets. Plato believed that astronomy was simply an extension of mathematics and that the movements of all heavenly objects could be explained by mathematical principles. Plato established a number of premises on the movement of the stars, sun, and moon. For example, he noted that the sun, moon, and planets in their movement from east to west followed a line in the sky known as the **ecliptic**. Furthermore, their orbits did not deviate more than eight degrees from the ecliptic. But once again, the planets and their erratic patterns of movement along the ecliptic presented difficulties in the construction of a simplified model of the universe.

The Greek philosopher Eudoxus was one of the first to apply Plato's concepts toward a working model of the universe. Although this idea is often credited to Aristotle, Eudoxus pictured the heavens as a series of shells, with each shell governing the movement of a specific object. These shells moved at a constant rate, as can be viewed by the movements of the stars at night. Eudoxus suggested that the sometimes erratic motion of the planets was because the planets were controlled by multiple spheres and the combination of their movement occasionally resulted in what appeared to be retrograde motion. While Eudoxus believed that the spheres were based solely on mathematics, Aristotle, with his interest in the mechanical nature of the universe, presented the idea that the heavens consisted of physical spheres, and were not merely mathematical representations. Aristotle constructed a complex network of spheres, over fifty in number, in an attempt to explain the astronomical observations of his time. Several centuries later, Ptolemy, using additional astronomical observations obtained by the Greek astronomer Hipparchus, would correct errors of Aristotle's earlier work. Ptolemy constructed a model of the universe that was centered on the Earth. Surrounding the Earth were crystalline shells on which rode the sun, moon, stars, and planets. The motions of these objects, specifically the planets, were governed by the movement of the spheres in relation to the earth. Ptolemy used his model to explain everything from the retrograde movement of the planets to the motions of the constellations. Although significant errors would accumulate in Ptolemy's model over the next centuries, it would remain the predominant model of the heavens for over fourteen centuries, one of the longest-lived scientific theories of record (see HELIOCENTRISM).

By the 16th century the accumulated errors with the Ptolemaic system had made the system almost unmanageable. While the constant motion

of the stars, sun, and moon fit into Ptolemy's system, the erratic motions of the planets were difficult to explain. A simpler model, one in
which the universe was not centered on the Earth, had been proposed
by the Polish astronomer Copernicus. In this model, the planets circled
the sun, not the Earth directly (see HELIOCENTRISM). Unfortunately, while this model presented a simpler and more logical explanation for centuries of astronomical observations, it did not fit well
with prevailing religious beliefs of an Earth-centered universe. Yet
change was inevitable. While observing comets, the skilled Danish astronomer Tycho Brahe noticed that comets followed elliptical orbits
and did not appear to be associated with any of the crystalline spheres
(see HALLEY'S COMET). In an attempt to preserve the Ptolemaic
model, Brahe devised a compromise system. In a hybrid between Ptolemaic and Copernican ideas, Brahe placed the planets in orbit around
the sun. The sun, in turn, orbited the Earth. While his model would
correct many of the existing problems in the Ptolemaic system, Tycho's
synthesis would be short-lived. Early in the 17th century, using a new
instrument called a telescope (see TELESCOPE), Galileo discovered
the moons of Jupiter (see JUPITER). Upon discovering that his identified moons of Jupiter were in orbit directly around the planet, Galileo
theorized that this system could serve as a model for the solar system.
In such a model the planets would orbit the sun in the same manner
as the Galilean moons orbited Jupiter. The strength of this idea effectively ended support for the Ptolemaic system.

In the early 17th century the new visions of space provided by the
first crude telescopes would spell an end to belief in crystalline spheres
orbiting the Earth. However, in the absence of spheres, there was a
need for a new theory of planetary motion. The pioneer behind this
transformation, and one of the first scientists to openly support a Copernican model, was the German astronomer Johannes Kepler. As Tycho Brahe's student, Kepler had access to the vast amounts of
observational data that Tycho had accumulated on the orbits of the
planets. Initially, Kepler was convinced that there was geometric harmony in the universe, and that the discovery of these geometric patterns would solve the problems of the Ptolemaic system. While this
initially led him to make incorrect assumptions on how the planets
were spaced in the solar system, it did force Kepler to address the role
of the sun in the solar system. More importantly, in an attempt to
confirm his ideas, Kepler was persistent in the belief that his hypotheses must be supported by physical evidence, a concept that was new
in the early 17th century.

Figure 12. Kepler's First Law of Planetary Motion. This states that all planets follow an elliptical orbit around the sun versus the circular orbit shown in the diagram. The sun serves as one focus for the orbit.

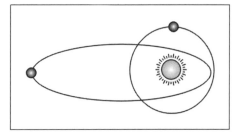

If the solar system were centered on the sun, then Kepler believed that it should be possible to calculate the Earth's orbit. The problem that Kepler faced was the difficulty in calculating the orbit of an object you are standing on without a reference point. Fortunately, Kepler had access to Tycho's precise observations on Mars. Using Mars as a reference point, Kepler was able to determine that the orbit of Earth was similar to the orbits of the other planets. However, when Kepler applied his new model of geometric harmony to the existing observational data on Mars, he noted that there existed an eight-degree error between his predictions and Tycho's observations. Clearly there was some other force at work, and Kepler believed that this force was directed outward from the sun. The theory of gravitation had yet to be established, but during Kepler's time there was an interest in magnetism as the predominant force of nature (see MAGNETISM). By performing elaborate calculations, Kepler concluded that the velocity of a planet is inversely proportional to the distance of the object from the sun. This would later become Kepler's Second Law of Planetary Motion.

These new data fit his earlier calculations on a circular orbit for the Earth, but still left an inadmissible level of error when applied to Mars. Before Kepler's time, astronomers attempted to correct these observed errors by assigning an epicycle to the orbits of the planets. An epicycle is a small circle that moves around the circumference of a larger circle, which in this case is the orbit of the planet. An epicycle may be thought of as a wobble in the orbit of a planet as it revolves around the sun. Historically, epicycles would be made as large or small as necessary to make the theory or idea fit the observational data. This was not acceptable for Kepler, so he once again turned his attention to the orbit of Mars. Kepler noted that if he made the orbit of Mars elliptical, rather than circular, the errors would be corrected. The elliptical, rather than circular, shape of planetary orbits later became the basis of Kepler's First Law of Planetary Motion (see Figure 12).

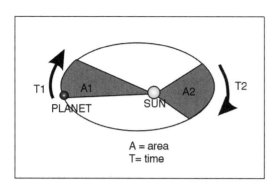

A = area
T = time

Figure 13. Kepler's Second Law of Planetary Motion. This states that an imaginary line stretched from the center of the sun to the center of a planet will sweep the same area in a given amount of time. Thus, planets farther from the sun move slower than planets closer to the sun. In this diagram A1 and A2 represent equal areas, while T1 and T2 represent equal elapsed times.

Armed with his new ideas, Kepler once again turned his attention to determining a geometric harmony in the universe. Kepler had strong religious beliefs and was convinced that the structure of the universe did not occur by chance. What Kepler was searching for was the harmonic law that justified his earlier discoveries. What he discovered, later to be called Kepler's Third Law of Planetary Motion, was that a ratio exists between the period of a planet's orbit around the sun and its distance from the sun (see Figure 13). Specifically, Kepler determined that the square of the period is proportional to the cube of the distance, or

$$t^2 = d^3$$

where t equals the period of revolution and d equals this distance from the sun (typically measured in astronomical units, or A.U.). While each of these laws was published independently throughout his career, Kepler summarized them in a book entitled *Epitome of Copernican Astronomy*. Prepared as a series of volumes and published between 1617 and 1621, this work became the basis of what later would be known as Keplerian astronomy. In his book Kepler outlined his three laws of planetary motion:

1. All planets follow an elliptical orbit around the sun. The sun serves as one focus for the orbit (Figure 12).
2. An imaginary line stretched from the center of the sun to the center of a planet will sweep the same area in a given amount of time. Thus, planets farther from the sun move slower than planets closer to the sun (Figure 13).

3. The squares of the period of the orbits of two planets are proportional to the cubes of their mean distances from the sun (see formula above).

Kepler recognized that while he had produced a functional model, one that presented a much simpler organizational pattern to the universe than Tycho's construction, he had not conclusively identified the force that kept the planets in their orbits. Kepler thought that a magnetic force was responsible for his observed laws. Later in the century, the English mathematician Isaac Newton developed a unifying theory on the structure of the universe. Newton compiled the ideas of Galileo, Kepler, and Copernicus around a single force, gravitation (see GRAVITATION). Presented as part of his epic work *Principia* (1687), Newton mathematically discussed Kepler's views on the motion of the planets. Instead of magnetism, Newton demonstrated that gravitation was responsible for the orbits of the planets. In doing so, Newton provided the needed proof of Kepler's laws of planetary motion.

Kepler's laws of planetary motion are significant for several reasons. While the laws themselves would eventually have an influence on the work on Isaac Newton, initially their primary significance would be to dispel the concept of an Earth-centered universe. When coupled with the discoveries of Galileo, Kepler's laws would leave little choice but to accept a Copernican model. Within a short period of time, Kepler and Galileo had uprooted an enduring foundation of Ptolemaic science. But initially what was more important was Kepler's method of investigation. Using a dogmatic approach to a problem, Kepler believed that scientific ideas and theories must be supported by available data. Kepler was forced to scrap his own idea of a circular orbit in favor of the elliptical orbits that his calculations supported. While the concept of scientific discoveries supported by facts may be commonplace in our century, at the dawn of the 17th century the idea was revolutionary. Kepler's laws of planetary motion were an important first step in the revolution of science.

Selected Bibliography

Evans, James. *The History and Practice of Ancient Astronomy.* New York: Oxford University Press, 1998.

Gingerich, Owen. *The Eye of Heaven: Ptolemy, Copernicus and Kepler.* New York: American Institute of Physics, 1993.

Kozhamthadam, Job. *The Discovery of Kepler's Laws: The Interaction of Science, Philosophy and Religion.* Notre Dame, IN: University of Notre Dame Press, 1994.

Wilson, Robert. *Astronomy through the Ages: The Story of the Human Attempt to Understand the Universe.* Princeton, NJ: Princeton University Press, 1997.

L

Law of Falling Bodies (ca. 1609–ca. 1632): There is a popular legend in science that Galileo derived his theories on the nature of gravity and falling bodies by dropping weighted objects from the top of the Tower of Pisa in Italy. While scientific historians have discredited the occurrence of this event, it does provide an indication of the direction that science was taking during the early 17th century. Since the time of the ancient Greeks scientists have attempted to explain the forces that control a falling object. As was typical of the times, the Greek philosopher Aristotle derived an explanation that was based solely on the power of observation, and not substantiated by experimental evidence. To Aristotle, the **acceleration** of an object after it was dropped was a function of the weight of the object, with heavier objects accelerating more quickly in comparison to lighter objects dropped from the same height. However, he did recognize that the density of the medium had an influence on the acceleration of an object. He considered this to be an inverse relationship. Thus, as the density of the medium through which the object was passing decreased, the speed increased. Unfortunately, Aristotle used this logic to support his idea that vacuums could not exist in nature, as by his definition an object in a vacuum would have infinite speed (see VACUUM PUMP). Although there were opponents of Aristotle's view, for example the Greek philosopher John Philoponus (ca. 6th century), who provided experimental evidence against Aristotle's ideas, his reputation and vastness of work in science and philosophy perpetuated his ideas until the time of the Renaissance.

In the opening years of the 17th century a number of physical forces, from planetary motion to magnetism (see KEPLER'S THEORY OF

PLANETARY MOTION; MAGNETISM), were under investigation. In each case the application of mathematics and scientific observation unveiled problems in prevailing thought. Investigations into the motion of falling bodies and projectiles are a prime example of this trend. There were many studies of motion during the 17th century (see LAWS OF MOTION), but the first to focus on the properties of falling bodies was the renowned Italian scientist Galileo. As with many of his discoveries, Galileo started with an observation and then asked how he could explain the observation in scientific and mathematical terms. In the case of falling bodies, Galileo had a number of preexisting observations with which to work. First, it was well known that an object at rest remains at rest unless acted upon by an outside force. This is called the Law of Inertia and would be investigated in considerable detail during the 17th century (see LAWS OF MOTION). Second, the concept of gravity as a force was recognized during Galileo's time, although its influence on motion was poorly understood (see GRAVITATION). Finally, and of the greatest importance, was an observation that the **velocity** of a falling object increases with time. Galileo used these ideas as the cornerstones for his developing ideas on falling bodies. This became one of the first organized scientific challenges to Aristotelian physics.

There were several important steps in the evolution of Galileo's Law of Falling Bodies that were necessary before Galileo could make a final presentation of his ideas in 1632. The prevailing concept of inertia prior to Galileo pertained primarily to objects at rest. Most believed that once an object was placed in motion it would eventually stop even in the absence of external resistance. Galileo was the first to formulate the idea that objects in motion continue in motion unless an external force is applied (see LAWS OF MOTION). He correctly deduced that the motion of an object was along a given **vector** under a constant velocity unless influenced by an outside force. Furthermore, these external forces were independent of the object; that is, the force was exerted whether the object was at rest or in motion. Galileo also separated the concept of velocity and acceleration. While many believed that falling objects exhibited a constant velocity, even though simple observations had suggested otherwise, Galileo proposed that it was the acceleration of the object that was a constant. Since the external force was applied throughout the duration of the motion, and not simply at the beginning, the effect of the force was compounded along the entire path of the object. What Galileo suggested was that the rate of change, or acceleration, from one moment to the next was constant and thus the velocity increased over the time of the fall. The distinction between velocity and acceleration was an important advance for the study of physical science.

As an integrator of mathematics and science, Galileo was not content with simply observing acceleration and velocity; he wanted to be able to predict the motion of a falling object. He proposed several mathematical proofs of his ideas on constant acceleration. As noted earlier, Galileo had established that the velocity (v) of a falling object at a given time was directly proportional to the amount of time (t) that it had been in motion. Using this information Galileo geometrically obtained the formula:

$$vt = 2s$$

where s represents the distance traveled. Since this is a direct relationship, if the object started at rest and accelerated to some velocity at the end of a specified amount of time, then the velocity would have increased at a uniform rate during the duration of the fall. Galileo then derived what is considered to be the Law of Falling Bodies. This is represented by the formula:

$$S = gt^2$$

The formula states that the distance traveled by a falling object, from rest to finish, varies with the square of the time that it is falling (g is a constant).

While these laws established the relationship between the variables, it was not based upon experimental evidence. Galileo wished to measure the rate of acceleration of a falling body. However, his ability to do this was limited by the lack of precision scientific instrumentation. For example, in the early 17th century there did not exist a reliable method of keeping time. While there were early models of the pendulum clock available, many of which were inspired directly by Galileo, these instruments were not capable of measuring time in small enough increments for acceleration studies. It was not until 1656 that Christiaan Huygens invented a clock that provided the needed level of precision (see PENDULUM CLOCK). Without access to a reliable instrument, Galileo needed to improvise in order to slow the acceleration down so that it could be measured using the timepieces available to him. To do this he measured the acceleration of objects as they moved down an inclined plane. Galileo's experiment was straightforward. He designed an inclined plane that had a small groove cut along its entire length. To reduce the amount of friction in the groove it was covered with smooth parchment. In addition, the height of the plane could be varied. A small brass ball was rolled down the entire length of the plane, and the time required to cover the distance was recorded. The ball was then rolled down exactly one-quarter of the distance and the time recorded. A comparison of the times revealed that the distance was proportional to the square of the time. In addition, Galileo noted that when the inclination of the plane was changed, the final velocity of the object remained the same. In other words, if an object initially was placed three feet above the ground and dropped, the final velocity was the same as if it was rolled down a thirty-degree plane elevated three feet from the ground. This provided the experimental proof for Galileo's Law of Falling Bodies.

Galileo's interest in motion was not limited to the study of falling bodies. Galileo remained concerned that his results were being influenced by air resistance, so he repeated many of the experiments using a pendulum. In doing so he defined the motion of a pendulum by noting that the period of a pendulum swing remained the same, while the distance the weight traveled decreased (see PENDULUM CLOCK). Galileo also applied these studies to the study of projectile motion, another area of physical science that had remained relatively unchanged since the time of Aristotle (see LAWS OF MOTION).

There can be no doubt that Galileo's study of the nature of falling bodies had a tremendous impact on the science of the 17th century. Not only did Galileo provide one of the first organized challenges to the aging and outdated scientific philosophy of Aristotle, but he firmly established the practice of linking mathematics to the study of natural phenomena. Although Galileo did not understand the nature of gravitation (such knowledge would have to wait until the work of Isaac Newton), he recognized that its effects could be expressed mathematically. From a scientific perspective this work led to the foundation of theories of inertia and motion, and without doubt influenced Newton in his study of gravitation (see GRAVITATION; LAWS OF MOTION). As later scientists worked to duplicate and expand on Galileo's studies of acceleration, there developed a need for more precise instrumentation, for example the invention of the pendulum clock (see PENDULUM CLOCK). Thus it can be said that for 17th-century science it was Galileo who first provided the link between scientific discovery and the invention of scientific instrumentation, a link that would remain strong well into the modern era.

Selected Bibliography

Cajorie, Florian. *A History of Physics*. New York: Dover Publications, 1962.

Drake, Stillman. *Galileo Studies, Personality, Tradition and Revolution*. Ann Arbor: University of Michigan Press, 1970.

Hall, A. Rupert. *From Galileo to Newton*. New York: Dover Publications, 1981.

Krebs, Robert. *Scientific Laws, Principles and Theories: A Reference Guide*. Westport, CT: Greenwood Press, 2000.

Lloyd, G. E. R. *Early Greek Science: Thales to Aristotle*. New York: W. W. Norton, 1970.

Laws of Motion (1632–1687): Since early times philosophers and scientists have studied the physical laws of the universe in an attempt to explain the nature of the world around them. In the area of physical science one of the longest-running areas of research has been directed toward understanding the forces responsible for the motion of an object. Motion is a complex physical action, one that involves many natural forces. Scientists have historically separated this into individual forces in order to understand them better before attempting to develop a unifying theory. These forces have included inertia, centrifugal force, and gravitation. Some of the earliest studies of motion were conducted on the nature of both falling bodies and projectiles. While the ancient Greeks began this investigation, it was not until the time of the Scientific Revolution that scientists possessed the necessary reasoning and experimental tools to effectively analyze the process of motion. In the 17th century these advances are best illustrated by examining the work of Galileo and Isaac Newton.

One of the first properties of motion that was investigated in the 17th century was the nature of falling bodies. The ancient Greek philosopher Aristotle believed that when an object was thrown into the air it fell to the Earth to return to its "natural" state (see CHEMISTRY). Furthermore, to Aristotle the weight of an object influenced its rate of fall, with heavier objects accelerating to a higher speed than lighter objects. To Aristotle, the force responsible for motion must remain in contact with the object while it is in motion. Thus, after a projectile was launched, its forward motion was due to air "pushing" it along its trajectory. The same held true for a falling object where the air "pushed" the object downward toward the Earth. He reasoned that this was because heavier items contained more of the element earth, which attracted them faster to the dense earth below them. Despite the obvious lack of experimental evidence supporting such an idea, it remained the prevalent thought until the 17th century. In 1632, this changed when Galileo formally presented his Law of Falling Bodies. This stated that the rate of acceleration of an object remained constant during a fall, although the velocity of the object increased (see LAW OF FALLING BODIES). While Galileo recognized that the uniform rate of acceleration was dependent on the force being applied, he did not understand the nature of the attractive force, although he believed that gravity was involved at some level. However, Galileo's definition of a physical force of nature in mathematical terms and the distinction he made between velocity and acceleration had a tremendous impact on studies of motion later in the century.

As part of his refinement of Aristotelian physics, Galileo also examined the motion of a projectile and the properties of inertia. Prior to Galileo it was well recognized that an external force was necessary to place an object in motion. However, since the concept of resistance to motion was not understood, it was widely accepted that after an object was placed in motion the velocity would gradually decrease until it stopped, even in the absence of a resisting force. Galileo recognized that once a body is placed into motion it would continue along a specific vector with a constant speed unless it interacted with a second force. However, the developing theories of heliocentrism presented a challenge for Galileo (see HELIOCENTRISM). To explain the motion of the heavens it was necessary to place the Earth in motion around its axis. Many questioned what the force was that kept objects from flying off of the Earth as it rotated, a problem that was not adequately solved until Newton's studies in gravitation (see GRAVITATION). Furthermore, many noted that if the Earth moved, then when an object was thrown into the air the Earth would move beneath it and thus the object should not return to its original location. But observational evidence suggested that this was not the case, since the object did return to its original location. In *Dialogue* (1630) Galileo explained that this was due to the inertia of the objects involved. If an object was dropped from the top of a building, as Galileo demonstrated in his Law of Falling Bodies, then the object fell to the foot of the building because the tower, the ball, and the Earth were all moving eastward with the same velocity. Thus the inertia of the objects maintained their relative position to one another. Opposition also arose when one examined the motion of an object when dropped from the mast of a moving ship. In this case the object often did not fall directly at the base of the mast. Galileo explained that while the ship, mast, and object were in motion, the air that the ship was passing through was stationary and thus represented a resisting force to the forward inertia of the object. Galileo then applied this concept to a centuries-old problem on the motion of a projectile. According to Aristotle's principles of motion, which were purely philosophical with no record of ever being tested experimentally, when an object is thrown into the air it should drop directly to the ground. As noted, for projectiles Aristotle believed that the air was pushing the object along. However, projectiles are observed to follow a parabolic curve. Once again Galileo turned to the subject of inertia for an explanation. The initial force that propelled the object provided the direction and velocity, while the curving of the object's path was the result of resistance from air and the influence of gravity. Galileo recognized gravity as a force, although it was not mathematically defined until Newton's work on universal gravitation (see GRAVITATION). Galileo's study of inertia was not only important in that it disproved the prevailing thoughts of Aristotle, but it provided the basis for Newton's later efforts on deriving the laws of motion. In fact, Galileo's inertia eventually evolved into Newton's First Law of Motion. This states that an object will continue along a given vector, or direction, with constant velocity unless acted upon by a second force.

In addition to his work with uniform acceleration and inertia, Galileo made some important contributions toward understanding the nature of centrifugal force and oscillations of a pendulum. When an object is in motion along a circular path, two forces are at work. The first is the attractive force that prevents the object from moving in a straight line according to the laws of inertia, while the second is an opposing force dependent on the momentum of the object. It should be noted that centrifugal force is not a true force under the Laws of Motion proposed later in the century by Newton, since if the inward force is removed, the object will return to a straight trajectory in response to inertia. It was Galileo who defined the concept of momentum by stating that it was a property of the velocity and weight of an object. This was later restated by Newton to indicate the **mass**, since weight is dependent on gravitational attractions. The German-American physicist Albert Einstein also redefined the properties of momentum as an object approaches the speed of light. However, both of these advances were based squarely on Galileo's work. For the motion of a pendulum Galileo noted that the period of the oscillation of a pendulum is a constant value no matter at what height one begins the swing. Later in the century, Christiaan Huygens continued Galileo's study of pendulum oscillations. This work resulted in the invention of the pendulum clock, an important time-keeping instrument for 17th-century science (see PENDULUM CLOCK).

In the late 17th century, Isaac Newton performed the final synthesis of these ideas into the Laws of Motion. In 1684 Newton published *De Motu* (*On Motion*), which expressed his theories on the motion of celestial objects mostly with respect to Kepler's earlier established laws of planetary motion (see KEPLER'S THEORY OF PLANETARY MOTION). This work would undergo a number of revisions, and eventually evolved into his epic work, *Principia* (1687). In *Principia*, Newton approached the laws of motion as a component of his work on the theory of universal gravitation (see GRAVITATION). He analyzed motion with respect to the forces involved, and expressed his proofs in mathematical notation. While Newton primarily used the tools of analytical geometry in his proofs, *Principia* introduced the calculus as a method of analyzing rates of change (see ANALYTICAL GEOMETRY; CALCULUS). Newton's first law of motion stated that an object at rest remains at rest unless acted upon by an outside force. This was a derivation of Galileo's work on inertia mentioned previously. The second law stated that change in the motion of an object in relation to its mass and velocity is proportional to the force involved in the change. This utilized Galileo's definition of momentum and defined the properties of centrifugal force. Finally, the third law stated that for every action there is an equal and opposite reaction. Using these laws it was possible for Newton to identify that in order for Kepler's laws of planetary motion to be true, the centrifugal force of the planets in their orbits must be countered by a second force that decreased with distance according to the inverse-square law. Newton demonstrated that this force was gravity (see GRAVITATION). Furthermore, he proposed that this force had universal properties and was responsible for the action of falling bodies, pendulums, and motions of planets and comets (see HALLEY'S COMET; LAW OF FALLING BODIES).

Newton's publication of *Principia* represents one of the most important scientific achievements in the history of science. With this work Newton not only summarized the achievements of the 17th-century scientists preceding him, but also synthesized their ideas into unifying theories on motion, optics, and gravitation (see GRAVITATION; OPTICS). In the study of motion, many of the scientists earlier in the 17th century had developed scientific theories of motion that lacked a final proof. For example, Galileo's Law of Falling Bodies recognized that external forces were involved, but did not classify gravitation as the force responsible for acceleration. A similar case was true for Kepler (see KEPLER'S THEORY OF PLANETARY MOTION). *Principia* not only validated the work of Galileo, Kepler, and Edmund Halley (see HALLEY'S COMET), but it did so by developing mathematical principles to explain the laws of motion. Prior to the 17th century the study of motion was based almost entirely on observation and philosophical discussions. Within a space of barely 100 years, scientists had not only designed experiments that refuted existing theories of motion, but explained their ideas as mathematical equations. The study of physical science would never be the same. Thus, the study of motion revolutionized science and represents one of the greatest scientific discoveries of the 17th century.

Selected Bibliography

Hall, A. Rupert. *From Galileo to Newton.* New York: Dover Publications, 1981.
Spangenburg, R., and D. K. Moser. *The History of Science from the Ancient Greeks to the Scientific Revolution.* New York: Facts on File, 1993.
Taton, Rene, ed. *The Beginnings of Modern Science.* New York: Basic Books, 1958.
Westfall, Richard. *The Life of Isaac Newton.* Cambridge: Cambridge University Press, 1993.

Light Spectrum (1665–1704): The study of light is a complex discipline that involves examining not only the physical structure of light (see LIGHT WAVES), but also how light interacts with other matter by the processes of **reflection**, **refraction**, and **diffraction**. Historically, the study of these properties has been called optics. Optical science has its roots in ancient civilizations including the Greeks and Arabs. However, modern investigations that apply mathematics and physical laws to the study of optics can be said to have originated around the time of the 17th century (see OPTICS). One area of research in optics that occurred during this time was the examination of the physical properties of color and the realization that light consists of a spectrum of components. Color in the natural world was obviously well known prior to the 17th century. The Greek philosopher Aristotle believed that all color was the result of a differential mixing of white and black. His student Theophrastus thought that the color of an object was due to its atomic configuration. In both cases there was little relationship between light and the color of an object.

This changed in the 17th century as researchers started to break apart beams of light. The Italian mathematician Francesca Grimaldi (1618–1663) performed one of the first experiments in this area. Grimaldi was primarily studying the refraction and reflection of light (see OPTICS). However, in one of his experiments (see LIGHT WAVES for figure), he noted that when he passed light through a small hole the edges of the light beam were broken down into bands of color. Furthermore, there was a blue band at one end of the beam and a red band at the other. Grimaldi used the results of this experiment to suggest that light behaved as a wave. He suggested that color was not a characteristic of an object, but rather was due to the wave motion of light interacting with the surface of the object by refraction, reflection, or diffraction. Thus, the color we see is actually that component of light that is reflected to us, while all the other colors (or **wavelengths**) are being absorbed by the object's pigments. This is the same basic principle as is recognized by modern science.

The English scientist Robert Hooke also contributed to the science of color in the 17th century. Hooke agreed with Grimaldi that color was not a property of an object, but rather a result of the interaction of light with an object. He designed several experiments to test his ideas. First, he discovered that if a beam of light were passed obliquely through a transparent object it would separate into red and blue beams. Hooke thought that light traveled faster in a denser object and thus the colors were produced when the leading edge of the beam accelerated once in contact with the object. Unfortunately, Hooke's ideas violated several of the existing laws regarding optics, namely Snell's Law (see OPTICS). However, like Grimaldi, he was correct in his overall assumption that color was a property of light.

Perhaps the most widely recognized experiments into the nature of color during the 17th century were the result of work performed by Isaac Newton. Unlike Grimaldi, Newton believed that light traveled as particles, and much of his research focused on determining the properties of these particles (see OPTICS). One of Newton's first experiments with color involved tying two colored threads together (red and blue) and placing them under a prism. When Newton looked at the threads through the prism the threads did not appear to be tied in a straight line. Instead, the red and blue threads were shifted in opposite directions. In a second experiment, if sunlight were passed through a prism, and the rays projected against a white surface (see Figure 14), a spectrum of colors appeared, with red and violet bands at opposite ends of the spectrum. From these two experiments Newton concluded that how much a prism deflects a ray of light is dependent on the

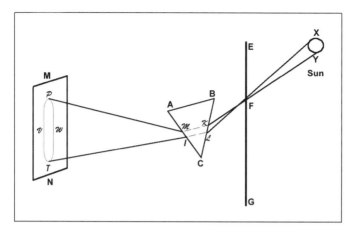

Figure 14. Newton's experiment (ca. 1672) that light consists of a spectrum of colors. The triangle ABC represents the prism with the area PT indicating the spectrum of colors. (Derived from *Philosophical Transactions* [1672].)

color of the light. In the experiment with the threads, the blue thread was deflected more than the red thread, which made the threads appear to be out of position.

Before Newton could conclude that color was an inherent property of light, he had to demonstrate that the color was not being created by the action of the prism. According to the scientific method, he had to construct a control experiment to eliminate the variable of the prism as a factor in color generation (see SCIENTIFIC REASONING). Newton constructed a simple, yet powerful, experiment to test his hypothesis that color was a property of light. In this experiment he used two prisms and two boards with small holes punched in them (see Figure 15). Light from the sun was passed through the first prism (ABC) and then the hole (G) in the first board (DE). The light coming through this first hole contained only one color of the spectrum, for example red. When this red light reached the second prism (abc), it was deflected intact as point M on the wall. In other words, the second prism did not add color to the beam and merely refracted the original color. In addition, if Newton adjusted the first prism (ABC) to send only blue light to the second prism, then the blue light was shifted to position N on the wall. Thus, as with the experiment with the threads, the color of the light determines how much the light is deflected by the prism. Furthermore, this experiment proved that the prisms were not responsible for the generation of the color. Newton also demonstrated that when a prism broke a light beam down into a spectrum of colors, these

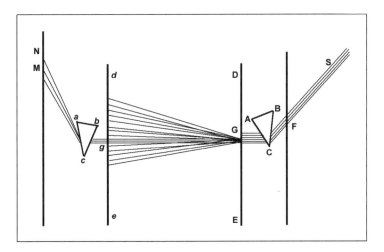

Figure 15. Newton's experiment demonstrating that color is an inherent property of light. See text for description. (Derived from *Philosophical Transactions* [1672].)

colors could be recombined back into white light by placing a second prism in the light path.

Newton had established that color was a property of light, but it remained possible that color was an inherent property of an object. To answer this problem Newton projected individual colors onto an object and observed its color. He noted that when the object was illuminated by a red ray of light, it appeared to have a red color. A red object would also appear to be red when exposed to a red light, but appeared to be dark when exposed to a blue light. Newton had determined that the color of an object was due to the colors that it reflected from its surface. Those colors that were not reflected were absorbed. The results of these experiments were initially published in Newton's first scientific work, *Philosophical Transactions* (1672), with updates appearing in both *Principia* (1687) and *Optiks* (1704). Within a relatively short period of time Newton had revolutionized the study of the spectrum of light with regard to color. While many questions were left unanswered, Newton had sent the physical science of optics down the experimental path to discovery.

Perhaps nowhere in nature is the spectrum of light more evident than when illustrated by a rainbow. As with the other discussions of color, scientific ideas on the nature of the rainbow date to the time of the ancient Greeks. Aristotle believed that rainbows were the result of the reflection of light from the underside of clouds. By the 14th century it was fairly well understood that the individual droplets of water were responsible for the creation of the rainbow, since they could be observed even in the absence of clouds. However, it was not until the 17th century when scientists mathematically developed the theories of reflection and refraction (see Snell's Law under OPTICS) that the contribution of the water droplet was better understood. Newton's theories of color played an important role in understanding rainbows. From his experiments it became obvious that the water droplets were serving as small prisms that were breaking up the incoming sunlight into colored rays. The angle of the sun against the droplets provided for the generation of the rainbow.

The work of these 17th-century scientists merely opened the door for the investigation of the light spectrum. The concepts of wavelength, visible spectrums and **absorption spectrums**, and **frequency** would take some time to develop, as did many other properties of light (see OPTICS). However, the accomplishments of Grimaldi and Newton were significant to the science of the times. Both had utilized the experimental approach to the understanding of a physical property of nature, a popular theme in late-17th-century science. In doing so, these scientists had demonstrated that the natural world could be explained logically. Newton's experiments are an integral component of many introductory physics courses where they clearly demonstrate not only generation of a light spectrum, but also the power of the scientific method. Subsequent generations of scientists would build on the principles and experiments in an attempt to more fully explain the nature of light. Studies of infrared and ultraviolet light, the power of lasers, and the spectrum of stars are all direct descendants of these initial studies.

Selected Bibliography

Park, David. *The Fire within the Eye: A Historical Essay on the Nature and Meaning of Light.* Princeton, NJ: Princeton University Press, 1997.

Spangenburg, R., and D. K. Moser. *The History of Science from the Ancient Greeks to the Scientific Revolution.* New York: Facts on File, 1993.

Wolf, A. *A History of Science, Technology and Philosophy in the 16th and 17th Centuries.* New York: Macmillan, 1968.

Light Speed (1600–1675): The velocity at which light travels has been a curiosity for scientists from the time of the ancient Greeks. Unlike sound, light appears to travel at an infinite speed. In a thunderstorm the lightning bolts are seen instantly, while the sound of the thunder takes time to reach the observer. While many early scientists accepted that light traveled at an infinite speed, the Greek philosopher Empedocles (ca. 495 B.C.E.) was one of the first to propose that light took time to travel through space, although even he believed that this speed was very great. However, like much of the science from his time, this theory was based on reasoning alone and not supported by scientific investigations. It was not until the advances in scientific thinking and instrumentation of the 17th century that honest attempts to ascertain the speed of light would be made possible.

Galileo, with his firm belief in the experimental approach, was one of the first 17th-century scientists to attempt to calculate the speed of light. Galileo's experiment was simple, and involved two observers signaling each other using lanterns. In this experiment the first observer would unveil his light and begin timing the experiment. Upon seeing the light of the first lantern, the second observer would respond by uncovering his light. The first observer could then measure the time that it took to send the signal and receive the reply. If one knew the response time of the second observer and the distance between the two points, it would theoretically be possible to calculate the speed of light. While based on sound principles, the effects of human error, coupled with the relatively short distances involved and a lack of precision time-keeping instruments, made an accurate calculation impossible. The investigation on the speed of light would first require advances in scientific fields that at first glance appeared to have little to do with the physical properties of light.

The invention of the telescope in the early 17th century encouraged the development of scientific instruments designed to make more precise measurements. Galileo's discovery of Jupiter's moons had moved observational astronomy to the forefront of science. The first instrument, initially designed to assist in **parallax** calculations (see MARS) and to determine the transit times of planets, was the pendulum clock. The pendulum clock served the scientific community of this century in many ways (see PENDULUM CLOCK). At the Royal Observatory of Paris, an important location for astronomical research during this time, the French astronomers Jean Picard and Giovanni Cassini were using the pendulum clock to make measurements on the orbits of the moons of Jupiter (see JUPITER). Specifically they were interested in timing exactly when each moon was eclipsed by Jupiter. The initial idea was to develop tables on the motion of Jupiter's moons and use this information in a similar manner as a lunar calendar. However, Cassini noted that when the Earth was close to Jupiter, the time when each moon was eclipsed by Jupiter became shorter, and when the Earth moved away from Jupiter, the time increased. The Danish astronomer Olaus Roemer, one of Cassini's assistants, theorized that it was taking the light from Jupiter longer to get to the Earth when the Earth and Jupiter were far apart, but without knowing the exact size of Earth's orbit, the calculation of the speed of light remained elusive.

Measuring astronomical distances had historically proven difficult for astronomers. However, the invention of the micrometer in 1666 (it had actually been invented earlier, but the invention had remained unknown until this time) by Adrien Auzout (1622–1691) made such calculations possible (see MICROMETER). Around 1667 Jean Picard used a micrometer to calculate the diameter of the Earth, an important factor for later calculations on the dimensions of the solar system. Johannes Kepler had made some preliminary calculations earlier on the orbits of the planets, Mars in particular, as part of his laws of planetary motion (see KEPLER'S THEORY OF PLANETARY MOTION). Once Picard had established the diameter of the Earth, it became possible to make more accurate distance calculations using parallax analysis. In 1673 Cassini and Jean Richer determined the parallax of Mars and using this information were able to calculate the orbit of Earth.

Roemer had previously noted that the difference in the timing of the eclipses of Jupiter's moons was twenty-two minutes. The actual time is sixteen minutes, but this is a reasonable estimate given the primitive nature of the instruments in use during that time. Roemer now had time (t) and distance (d) and thus could calculate the rate (r) of movement using the standard rate-distance formula

$$rt = d \text{ or } d/t = r$$

Roemer's calculations gave the speed of light as 140,000 miles per second. While this is not entirely accurate, as the actual value is approximately 186,000 miles per second, it did give an important first estimate of the velocity of light. In the next century, this work was continued by the English astronomer James Bradley (1693–1762). Using improved instrumentation, Bradley was able to improve the estimates of the speed of light to about 176,000 miles per second. In 1849, the French physicist Armand Fizeau (1819–1896) revisited Galileo's idea of timing light between two points and in doing so calculated a speed of 196,000 miles per second. Later experiments by the French physicist Jean Foucault (1819–1868) and the American physicist Albert Michelson (1852–1931) yielded values even closer to the current value.

Roemer's manuscript on the velocity of light was published in 1675 and quickly caught the attention of the scientific community. When coupled with the dimensions of the solar system worked out by Cassini and others, the universe was fast becoming an immense expanse of space, and the importance of the Earth was diminished. However, discussion on the velocity of light would not end in the 17th century with Roemer's work. The question could now be raised as to whether light traveled at different speeds in different mediums, in a manner similar to sound in water and air. Christiaan Huygens used this concept to support his theories that light existed as a series of waves that pushed through the media, an idea that was actively refuted by Edmund Halley and Isaac Newton (see LIGHT WAVES).

These early investigations on the velocity of light, and the principles behind that velocity, were important achievements for the scientists of this century. Science had entered the century having changed little since the time of the ancient Greeks. The scientific revolution of the 17th century not only introduced experimentation as a method of validating scientific ideas, but also marked the linking of technology and science. Technological advances in scientific instrumentation were making more detailed experiments possible, and allowed scientists to ask increasingly more detailed questions on the nature of the world around them. An excellent example of this is the 20th-century presentation of the Theory of Relativity by the German-American physicist Albert Einstein, which in part discusses the speed of light in a vacuum. This discovery is one of the most important in 20th-century physics, and is in part built on the foundation of knowledge that began in the 17th century.

Selected Bibliography

Asimov, Isaac. *Eyes on the Universe: A History of the Telescope.* Boston: Houghton Mifflin, 1975.
Park, David. *The Fire within the Eye: A Historical Essay on the Nature and Meaning of Light.* Princeton, NJ: Princeton University Press, 1997.

Light Waves (1604–1704): Representing one of the more visible forces of nature, the study of the physical properties of light was a popular area of research in the 17th century. Using improved instrumentation and reasoning skills, scientists of this time studied light from a variety of perspectives. During the 17th century there were significant advances in the understanding of the speed of light and the processes of reflection and refraction (see LIGHT SPEED; OPTICS). The developments during this century were based on the scholarship of ancient Greek and Arab scientists over the preceding two millennia. In the 3rd century B.C.E. the Greek mathematician Euclid reviewed the known research on the properties of light. His book *Optics* remained the authority on optical science until the 17th century. It was well established during Euclid's time that light moves in a straight line. This can be demonstrated experimentally since one can interrupt that path of light by simply placing an object between the source and one's eyes. Euclid also knew of the properties of refraction and reflection, although he was one of the first to attempt to explain them from a mathematical perspective (see OPTICS). What remained unclear was how the light moved from the source to the eye and furthermore how an image was detected. Some contended that particles were emitted from the eye, and then reflected from the object. Others suggested that the media between the object and the eye (air, water, etc.) was responsible for the transmission of the image. In the 11th century, considered to be the Dark Ages of Western science, the Arabic scholar Ibn al-Haytham (965–1039, also known as Alhazen) further advanced the study of optics with his work on reflection and refraction (see OPTICS). This work included a brief discussion on light waves in which he proposed that light moves outward from its source in a spherical pattern. However, the true nature of light waves remained elusive until the late 17th century.

In the 1600s the German astronomer Johannes Kepler made one of the first discoveries regarding the movement of light. Kepler's primary interest was in the motion of the planets and how that related to the structure of the solar system (see KEPLER'S THEORY OF PLANETARY MOTION). In these studies Kepler focused on describing the force that holds the planets in their orbits. Although he believed this to be magnetism, he did note that the strength of this force decreased with distance (see MAGNETISM). The English physicist Isaac Newton later used this relationship, called the inverse-square law, in his development of a unifying theory of gravitation (see GRAVITATION). However, Kepler also applied this relationship to the light waves. He commented that the intensity of light decreases proportionally with the distance from the source. Mathematically, as distance increases the intensity decreases at a rate proportional to the square of the distance. The

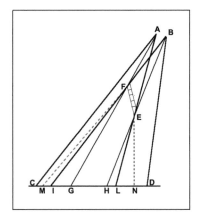

Figure 16. Grimaldi's experiments indicating the wave nature of light waves. See text for description.

French philosopher René Descartes expanded on Kepler's views during his own investigations of reflection and refraction (see OPTICS). To Descartes, the universe operated as a machine, with each small component contributing to the operation of the larger system. Since light traveled in some manner from its source, it must be able to be explained using laws of motion. Descartes believed that what we viewed as light was actually the motion of the medium, and not the light itself, between the source and our eye. Thus to Descartes, the motion of light was enhanced in a dense medium, and inhibited in a medium with a low density of particles.

While the contributions of both Kepler and Descartes introduced some important theories on the motion of light, neither truly addressed the physical properties of light motion. As the 17th century approached, two theories were gaining popularity. First was that light leaving the source moved as a series of waves through the medium. These waves are similar in nature to the waves that result from throwing a stone into a lake. The second possibility was that light itself was a particle that was emitted from its source and traveled until it was detected. One of the first distinctions between these two theories was made by the Italian physicist Francesco Grimaldi. As with many other investigators of the time, Grimaldi's research primarily focused on the reflective and refractive properties of light, although this also led to some useful insights into the motion of light waves. Grimaldi designed an experiment (see Figure 16) in which a beam of light was passed through a small hole (defined by points A and B) onto a white surface (represented by the line connecting points C and D). An opaque object (E and F) interrupted a portion of the path of the light. Before the object was in place, the light from point A illuminated a triangle represented by the points ACL. The light from point B was defined by the triangle BID. Assuming that light travels in a straight line, when object EF is placed into position, the area represented by the line GL should be in shadow owing to the interruption of light from source A. The same would be true of the area IH for the light coming from source B. This should leave the area GH completely dark. Grimaldi discovered that his research did not match these predictions. Instead, Grimaldi discovered that the shadowed area was larger than predicted and extended outward to points M and N. Grimaldi's explanation for this was that light actually moved through the medium between the source points and the paper as a wave. The deviations from the predicted outcome were due to ripples on the light wave that extended outward from the normally straight path of the light. This was one of the first experiments that described light in comparison to a wave. Later, Isaac Newton would duplicate Grimaldi's experiments, but conclude that results could be explained better if light consisted of particles. Grimaldi also used this experiment to investigate the spectrum of light as it related to color (see LIGHT SPECTRUM).

In 1676 the astronomer Olaus Roemer mathematically determined the speed of light to be about 140,000 miles per second (see LIGHT SPEED). While this value was lower than the actual value, it did serve to finally refute the idea that speed of light was infinite. Two questions arose from this experiment. The first was whether the speed of light varied with the type of medium. The second question once again focused on the characteristics of light's movement. Preliminary answers to these questions first appeared in *Treatise on Light* (1690) written by the Dutch scientist-mathematician Christiaan Huygens. In this work Huygens clearly presents the idea that light is matter, and thus may be explained as a mechanical process. Huygens contended that light moved in a similar wave-like motion as sound, and thus was propagated by the medium in which it traveled. Huygens opposed the idea that light was made up of small particles, suggesting that if this were the case then the particles would be in a constant state of collision with one another. The problem was that if one placed an object in a glass jar and removed the air using a vacuum pump (see VACUUM PUMP), it was still possible to see the object. This was not possible if the medium was responsible for moving the light. Huygens returned to Aristotle's idea of the elements air and ether (see CHEMISTRY). He suggested that light moved not through air, but through the **ether**. Huygens then explained how these waves originate from the source as a series of waves. Since there are countless waves originating from the source, and our eyes can't distinguish each individual wave, the image appears to be clear.

The final chapter in the investigation of light waves during the 17th century was conducted by Isaac Newton. Newton summarized much of the work in 17th-century optics in his 1704 publication entitled *Opticks*. *Opticks* contains some of the most detailed discussions to date on the nature of the light spectrum as it relates to color. Newton demonstrated that light consisted of discrete parts that could be separated and analyzed (see LIGHT SPECTRUM). He knew that light basically moves in straight lines, although Grimaldi's experiment suggested a wave-like nature. What he suggested was that light was actually a series of particles, and that color was the result of these particles striking an object and creating a vibration within the matter of the object. This theory of particles is frequently called the corpuscular theory. The particle nature of light had been proposed a few years earlier by Robert Hooke. However, it was Newton who made detailed descriptions on the origin of these particles. Newton does not seem to discount the idea of light moving as a wave, although he didn't determine how this relates to the obvious observation that light moves in a straight line. He suggested that light might originate as a wave, as proposed by Huygens, but the individual particles leaving the light source travel in a straight line until they interact with other matter. Newton relied heavily on the particle nature of light in order to explain the basis of color (see LIGHT SPECTRUM) and how light is refracted from an object. His experiments with color also led Newton to believe that the motion of light was cyclical, a phenomenon called periodicity. However, Newton believed that this was associated with the oscillations of the particles during reflection, and not a function of the true wave motion of light waves. Furthermore, his ideas supplemented experiments in the late 17th century on reflection and refraction, specifically those conducted by Descartes and Snell (see OPTICS).

The debate over the nature of light as a wave, a particle, or a combination of the two continues in modern science. However, both Newton and Huygens were partially correct. Light waves behave both as waves according to Huygen's ideas, and as particles as set forth by Newton. The movement of light in a vacuum, believed by 17th-century scientists to be possible owing to the presence of a compound called ether (see VACUUM PUMP), became a focus of several 19th-century studies. By the late 19th century most physicists did not accept the existence of ether and proposed other mechanisms, such as electromagnetism, for the movement of light. The wave and particle nature of light waves continues to be investigated in modern science, and despite the sophistication of modern methods, a number of important questions still remain. However, the basic principles on which modern physicists investigate light waves is founded on the work of these 17th-century pioneers.

Selected Bibliography

Park, David. *The Fire within the Eye: A Historical Essay on the Nature and Meaning of Light.* Princeton, NJ: Princeton University Press, 1997.
Wolf, A. *A History of Science, Technology and Philosophy in the 16th and 17th Centuries.* New York: Macmillan, 1968.

Logarithms (1614–1628): The scientific revolution of the late 16th and early 17th centuries, with its focus on investigation by experimentation, required a corresponding change in the field of mathematics. While some progress had been made in the preceding centuries in the study of exponents and progressions, it was not until the invention of logarithms early in the 17th century that an accurate means of handling large number calculations would be possible. To understand logarithms it is first necessary to be familiar with the term *progression*. A mathematical progression is best described as a set of numbers, or quantities, in which each member of the set differs from the preceding value by a set amount or constant. In an arithmetic progression the sequence is established by adding a constant to each quantity. For example, the sequence 1, 3, 5, 7, 9 is an arithmetic progression with a constant of 2. A geometric series is developed along a similar concept except that the constant value is multiplied to obtain the next value in the series. In the geometric progression 1, 2, 4, 8, 16 the constant value is again 2, but this time the progression increases by a *factor* of 2. In the geometric progression below, the constant value is x.

$$1, x, x^2, x^3, x^4, x^5$$

Since $x^0 = 1$ and $x^1 = x$, an examination of the exponents yields the arithmetic progression 0, 1, 2, 3, 4, 5. If the number 10 is substituted for x, then if 10^3 is multiplied by 10^2 the resulting value is 100,000, or 10^5. Thus the product of the two numbers is equal to the base number with an exponent equal to the sum of the two exponents. Similarly, when dividing two exponential numbers with the same base, the product has an exponent equal to the difference between the exponents. These laws of exponential notation had been worked out in the 15th century by Nicolas Chuquet (ca. 1445–ca. 1500). The major drawback of exponential notation it is that it most useful when multiplying or dividing numbers with whole number exponents. For example, the numbers 32 and 64 are easily represented by 2^5 and 2^6, respectively. However, any value between 32 and 64 would be represented by some fraction of a value between 5 and 6.

In the later half of the 16th century there was a renewed interest in mathematical astronomy, specifically with the use of **trigonometry**. Trigonometric calculations involving both very large and very small numbers had presented special problems for scientists of this time. A number of individuals simultaneously attempted to develop a method to expedite these cumbersome calculations. One such individual was the Scottish mathematician John Napier. Napier was not directly interested in astronomical calculations, although he may have shared some of his work with the 16th-century astronomer Tycho Brahe. At the time Brahe was performing complex calculations in an attempt to correct the antiquated Earth-centered model of the solar system that had persisted since the time of the ancient Greeks (see JUPITER; KEPLER'S THEORY OF PLANETARY MOTION). Napier recognized that by using trigonometric principles it could be possible to calculate an exponential value for any number at a given base. Napier adopted the term *logarithm*, from the Greek words for number and ratio, for this type of calculation. At about the same time, the Swiss mathematician Joost Bürgi (1552–1632) had developed a similar logarithmic method. However, in 1614 Napier would be the first to publish an explanation of logarithms. This was followed by a second publication in 1619. Thus, credit for the invention of logarithms is most often bestowed upon Napier. Once again, the simultaneous discovery of the same idea is a strong indicator of the need for the invention or discovery by scientists in a given period of time. This was typically because science had reached an obstacle that required an innovative approach in order for it to be overcome.

Logarithms may be expressed in the form:

$$\log_2 16 = 4$$

This notation indicates that 2^4 equals 16, or in other words, that 4 is the logarithm of 16 in base 2. Note that the logarithm of a number is dependent upon the base being used. While the logarithm of 16 in a base 2 system is 4, in a base 4 system the logarithm would be 2 ($4^2 =$

16). However, since we typically operate in a base 10 system, bases other than 10 can be difficult to understand. After the publication of his work in 1614, Napier consulted with the English mathematician Henry Briggs (1561–1630). Briggs suggested that they focus on using a base 10 system for logarithms, an idea that persists to this day. Logarithms using a base 10 are often called Briggsian, or common, logarithms in acknowledgment of this contribution. By convention, base 10 logarithmic equations do not indicate the base, for example:

$$\log 100 = 2$$

Since the goal was to develop logarithmic tables that would ease calculations, Napier and Briggs also decided to set the scale of logarithms so that $\log 1 = 0$ ($10^0 = 1$) and $\log 10 = 1$ ($10^1 = 1$), an idea consistent with the modern use of exponents. However, it should be noted that during the early 17th century the use of fractional exponents had not yet been accepted. The link between logarithms and exponential notation would have to wait until the English mathematician William Jones (1675–1749) introduced the idea in 1742.

Once the process of calculating logarithms had been established, it was then possible to begin construction of logarithmic tables. Logarithms are nonrepeating decimals. For example:

$$\log 78 = 1.892094603$$

Thus, as the number of digits in the logarithm increases, so does the accuracy of the calculation. For most logarithmic calculations, five decimal places provides an adequate level of accuracy, although expanded tables exist for more precise measurements. In 1617 Briggs published the first logarithmic tables for the numbers between 1 and 1,000. These values were calculated to fourteen decimal places, an amazing effort considering it was done without the aid of modern calculating devices. In 1624, Briggs extended the tables to include the numbers between 1 and 20,000 and 90,000 to 100,000, again to 14 decimal places. The Dutch mathematician Adriaan Vlacq (1600–ca. 1666) contributed the missing interval in 1628. Thus over the span of a few decades mathematicians had succeeded in developing a table of logarithms that remained in use for the next 350 years until the invention of the modern calculator.

Logarithms quickly found widespread application in many areas of science. In 1620 Edmund Gunter applied logarithms to trigonometric functions of sine and tangent and in doing so, introduced the terms *cosine* and *cotangent*. Gunter also collaborated with William Oughtred in creating the first mechanical device to calculate logarithms, the slide rule (see SLIDE RULE). The 18th century would see an expansion of logarithmic theory by the Bernoulli family to include negative logarithms and the expression of logarithms in exponential form. The use of logarithms was not limited to mathematics and astronomy however. For example, in the biological sciences a relationship exists between the concentration of hydrogen ions in a solution and the acidity of the solution. This is best expressed by the equation:

$$pH = \text{-log } [H^+]$$

In other words, as the hydrogen ion concentration changes by a factor of ten, there is a corresponding one-point change in the pH of the solution.

The invention of logarithms at the start of the 17th century had a significant impact on the science of the times. Johannes Kepler's calculations on the orbits of the planets, which would eventually lead to his laws of planetary motion (see KEPLER'S THEORY OF PLANETARY MOTION), were greatly aided by the use of logarithms. While mathematical theory had made a few advancements prior to the start of the 17th century, the invention and use of logarithms marked a revolution in mathematics. With the intense interest in describing the mathematical laws of the universe, such as Isaac Newton's work on gravitation (see GRAVITATION), there would come a need for more complex mathematical theories. The logarithms described by Napier and Briggs formed the basis of these, and future, complex calculations. Logarithms would remain a standard in mathematics until the invention of the modern calculator in the 20th century (see CALCULATING MACHINES).

Selected Bibliography

Boyer, Carl B., and Uta C. Merzbach. *A History of Mathematics.* 2nd ed. New York: John Wiley & Sons, 1989.

Cooke, Robert. *The History of Mathematics: A Brief Course.* New York: John Wiley & Sons, 1997.

Katz, Victor J. *A History of Mathematics: An Introduction.* Reading, MA: Addison-Wesley Longman, 1998.

Swetz, Frank J. *From Five Fingers to Infinity: A Journey through the History of Mathematics.* Chicago: Open Court Publishing, 1994.

Lymphatic System (1622–1656): The majority of the scientific advances made during the 17th century were the result of a new era of science that coupled experimentation and innovation in the scientific method. However, this was not the case for all disciplines. In the case of medicine, specifically the study of human anatomy and disease, physicians and scientists were limited in their experimental methods and thus remained confined to science by observation alone. Although some progress was being made, as may be witnessed in William Harvey's studies on the circulation of the blood (see BLOOD CIRCULATION), the study of human anatomy had changed little since the time of the ancient Greeks. Most of the descriptions and theories of human anatomy were established by the Greek physician Galen in the 2nd century. These theories remained relatively intact and unchallenged for almost 1,400 years. Studies of circulation in Western culture date back to this time and while Galen developed detailed theories on the physiology of human blood circulation, missing was the role of the lymphatic system. This is despite the fact that several centuries earlier the Greek anatomist Erasistratus (ca. 304–ca. 250 B.C.E.) had described its presence. Subsequent studies, including those by the 16th-century Belgian scientist Andreas Vesalius, also failed to mention the role of the lymphatic system in circulation (see BLOOD CIRCULATION). While in modern medicine the lymphatic system is recognized as a major circulatory system of the body, historically its function has probably been confused with that of the blood circulatory system. The lymphatic system has several important functions, including the transportation of fat-soluble nutrients and regulation of water levels in tissues as well as serving as a conduit for components of the immune system. During the 17th century a number of observations were made that began to identify the individual components of this system. This would set the stage for more physiological studies over the next two centuries.

The Italian physician Gaspare Aselli (1581–1625) made the first written accounts of the lymphatic system in Western medicine. Around 1622, while studying the digestive system of a dog, Aselli noted that the lymph glands contained a cloudy fluid. As a supporter of Galen's model of circulation, Aselli incorrectly concluded that these glands were somehow associated with blood circulation and that the contents were destined for the liver. However, these structures, also called the *lacteal vessels*, are now recognized as a component of the lymphatic system and serve to remove absorbed fats from the digestive tract. Aselli's accounts of the lymphatic system also represented another first for medical science in that he published the diagrams of his discovery in color. This work served as an inspiration for a number of other physicians who strove to identify additional examples of Aselli's vessels, commonly called lacteals. One of these was the French physician Jean Pecquet (1622–1674), who noted that in the dog a set of lacteals in the upper body are drained by a reservoir. This reservoir is now called the cisterna chyli or receptaculum. Pecquet also discovered that the lymph from this structure exits by means of the thoracic duct, eventually to be deposited into the circulatory system.

The Danish physician Thomas Bartholin (1616–1680) was the first to apply Pecquet's studies to humans. In 1652 he published *De Lacteis Thoracsis* in which he describes the anatomy of the thoracic duct in humans. It was Bartholin who first recognized that these lacteal vessels formed a second circulatory system within the body. He named this system the *lymphatic system*. The Swedish physician Olof Rudbeck (1630–1702) provided a more extensive account of the lymphatic system. Around 1653 Rudbeck published his studies of the lymphatic system that were allegedly the results of his studies in hundreds of animal species. More importantly, Rudbeck finally disproved the connection between the lymphatic system and the circulation of the blood when he showed that the lymphatic fluid does not flow toward the liver, but rather is a product of the liver. The English physician Francis Glisson (ca. 1597–1677) made an early prediction of the function of the lymph fluid when he suggested that it had some role in the lubrication of the major cavities of the body. Unfortunately he provided no proof of this idea and it did not seem to have an impact on the study of the lymphatic system during his time. Finally, the Italian microscopist Marcello Malpighi noted that the lymph glands, whose widespread presence in the human body had been detected by a number of investigators, were a component of the lymphatic system and not the fibrous growths that many believed. He also demonstrated that the lymphatic fluid moved through these structures.

Some accounts credit the discovery of the lymphatic system to two 18th-century physicians, William Hunter (1718–1783) and Alexander Monro (1733–1817). Hunter and Monro independently developed the absorbent nature of the lymphatic system. This basically stated that the function of these interconnected vessels was to drain fluids from the tissues and deposit them into the circulatory system by way of the thoracic duct and jugular vein. While this idea was suggested first by Glisson, the correctness of his statements was not realized until the 18th century. In conclusion, the discoveries of the mentioned 17th-century anatomists and physicians served two functions. First, they provided additional evidence against Galen's concept of the circulatory system, and second, the identification of lymphatic glands and vessels served as an important foundation for the development of a later unified theory on the role of the lymphatic system in human physiology.

Selected Bibliography

Duffin, Jacalyn. *History of Medicine: A Scandalously Short Introduction.* Toronto: University of Toronto Press, 1999.

Eales, Nellie B. "The History of the Lymphatic System, with Special Reference to the Hunter-Munro Controversy." *Journal of the History of Medicine and Allied Sciences* 29, no. 3 (1974): 280–294.

M

Magnetic Declination (1600–ca. 1692): Magnetism has long been recognized as an important force of nature. Historically, magnetism was known to the ancient Greeks as early as the 7th century B.C.E. Around that time the Greek philosopher Thales made some descriptions of the properties of lodestone, a magnetic ore. However, it was the ancient Chinese who made some of the first technological advances associated with magnetism when they recognized that lodestone would orient itself in a north-south configuration. Some believe that the 13th-century Italian explorer Marco Polo (ca. 1254–1324) was responsible for bringing the Chinese lodestone compass back to Western culture. In the time between these cultures and the 17th century magnetism was primarily associated with the practice of medicine, where it was believed to help cure certain ailments. In the 17th century the English physician William Gilbert conducted the first true scientific studies of magnetism. Not only did Gilbert recognize magnetism as a force of nature, he also distinguished its attractive force from that of electricity (see MAGNETISM). From a historical perspective Gilbert's work is frequently recognized as one of the first in the scientific revolution of the Renaissance to employ the scientific method (see SCIENTIFIC REASONING) and as such had a tremendous influence on the science of the 17th century.

Prior to the 17th century sailors had discovered that the needle of a magnetic compass does not point directly north, but typically displays some small degree of error. This error was commonly called the *variation of the compass*, or magnetic declination. Furthermore, that degree of error was not a constant, and was dependent on the location where the reading was being made. This presented a special problem for maritime commerce and explorers who typically relied upon longitude and latitude to determine their position at sea. Even small variations in position can produce large navigational errors over long distances. As is often the case, the technological needs of society directed the path of scientific advancement. By the 17th century several instruments had been designed to measure the magnetic declination. These instruments typically used the sun and stars as fixed reference points and then measured the difference between true north and the magnetic north being displayed by the compass. During the 16th century there was an attempt to map magnetic declination around the globe. This was done by measuring the angle of variation as well as measuring the dip of the compass needle at a given location. There was hope that the amount of magnetic declination would follow a measurable pattern on the ocean and thus could be used as a navigational aid for sailors. Unfortunately, by the 17th century it was recognized that the magnetic declination not only varies with the location, but also changes over time.

Gilbert's study of magnetism provided the first understanding of the nature of this phenomenon. Gilbert believed that the Earth behaved in a similar fashion as a lodestone magnet. When Gilbert passed a compass over the surface of a magnet, he found that the amount of dip in the needle varied with the distance of the compass from the poles of the magnet (see MAGNETISM). The dip was greatest at the poles, and least at the equator of the magnet. Gilbert recognized the similarities between his model and the observed maritime measurements. He attributed the deviations in the amount of magnetic declination at a specific location to the fact that the surface of the Earth was not smooth and the variations in its surface influenced the action of the compass. Later he proposed that this field extended outward from the surface of the planet (see MAGNETISM). In addition, his limited data suggested that the needle of a compass was deflected toward land masses and away from large open bodies of water. Gilbert attributed this to the nonmagnetic nature of water and the presence of magnetic ore under land. Using his lodestone magnet as a model, Gilbert demonstrated that by quantifying the dip in the needle, it was possible to determine how far one was from the magnetic poles. These ideas were presented in *De Magnete* (1600), a publication that would have a strong influence on the work of Johannes Kepler several decades later (see KEPLER'S THEORY OF PLANETARY MOTION). *De*

Magnete was an important first step in this field, but it did present some incorrect views, namely the fact that the geographic and magnetic poles of the Earth were the same (see MAGNETISM).

Later in the 17th century research in magnetic declination focused on explaining why the amount of declination varied with time. Earlier in the century a number of scientists had observed that at a given location the variance from true north varied by degrees even over a span of a few decades. Some suggested that this was due to the mining or oxidizing of iron ore, but the global level of variation did not support this hypothesis. The leading researcher in the search for an explanation was the English astronomer Edmund Halley. Using data that he collected from expeditions into the Atlantic Ocean, as well as from other sources, Halley disproved the idea that it was the iron ore deposits that caused the variation. He also demonstrated that Gilbert's view of declination toward land masses was incorrect as there were areas in which the needle deflected toward the open sea and away from land. Instead he demonstrated that the magnetic poles are in a constant state of motion and that at any given moment areas of equal magnetic declination existed on the planet. From this Halley concluded that in addition to geographic poles, the Earth possesses four magnetic poles, two of which are located in each hemisphere. These poles are in a constant state of motion with a specific period of rotation. Using this information, Halley recognized that during a specific period of time it was possible to map the level of magnetic declination on the globe, although the values would drift in subsequent years.

Halley's ideas of the magnetic nature of the planet, founded on the pioneering work of Gilbert, were very close to modern theories on the Earth's magnetism. The Earth's magnetic poles and magnetic field are recognized to be in a constant state of motion. The magnetic poles of the North and South Hemisphere are also recognized to reverse polarity over geological time. Furthermore, the magnetic activity of the sun, as illustrated by sunspots, is also known to have an influence on the projection of the Earth's magnetic field into space. In the 17th century the study of magnetic declination was important in that it provided a testing ground for developing theories of magnetism. Magnetism was initially believed to be the force that was responsible for the structure of the universe (see KEPLER'S THEORY OF PLANETARY MOTION; MAGNETISM), at least until the description of gravity as the unifying force (see GRAVITATION). Thus, the study of its properties had a strong influence on the 17th-century scientists who sought to describe the physical forces of nature.

Selected Bibliography

Ronan, Colin. *Science: Its History and Development among the World's Cultures.* New York: Facts on File, 1982.

Taton, Rene, ed. *The Beginnings of Modern Science.* New York: Basic Books, 1958.

Wolf, A. *A History of Science, Technology and Philosophy in the 16th and 17th Centuries.* New York: Macmillan, 1968.

Magnetism (1600–1641): The concept that certain materials have magnetic capabilities, or the ability to attract other materials, predates the 17th century. Ancient Greek philosophers such as Thales provided some of the first descriptions of the attraction characteristics of lodestone, a naturally occurring iron oxide ore. However, it is more likely that the Chinese were the first to apply these attractive forces to the needs of their society. They may have been the first to notice that a freely rotating piece of lodestone would orient itself in a north-south direction. It is conceivable that the Chinese may have passed this information to the Europeans through the Arabic cultures, possibly with the aid of the Italian explorer Marco Polo, although the exact path has not been well established. By the 13th century studies of magnetism had again returned to Europe. The French scholar Peter Peregrinus (ca. 1240) invented crude lodestone compasses that were capable of giving directional information (north, south, east, and west). This led to the use of the directional terms *north* and *south pole* to indicate the orientation of the compass. Later, the use of the compass would lead to discoveries of variations in magnetic fields (see MAGNETIC DECLINATION). However, it would be almost four centuries before scientists would attempt to mathematically explain magnetism as a natural force of nature.

The dawn of the 17th century saw a resurgence in interest on the properties of magnets. Leading these investigations was an English physician named William Gilbert. Gilbert was not only a pioneer in the study of magnetism, but he also helped to establish the experimental method of scientific observation. Gilbert was a proponent of the belief that scientific theories should be supported by observations and verifiable by experimental procedures. This was the initial stage in the development of the scientific method. In his experiments with magnetism, Gilbert is recognized as one of the first scientists to apply mathematical principles to scientific observations. This change in philosophy would have a significant impact on the development of scientific thought in the remainder of the century.

William Gilbert appears not only to have been aware of, but actually continued, the work begun on magnetism by Peregrinus. These early works were complicated by a developing interest in other attractive forces, such as electricity. Often, attractive forces were all combined as one general physical force of nature. Gilbert was able to clearly distinguish between the attractive forces of magnets, which were confined to iron-containing compounds (such as lodestone), and other forces such as electricity. As a physician, Gilbert was interested in the medicinal properties of magnets. Magnets were believed to be capable of treating a wide assortment of illnesses, from headaches to bowel problems. In his writings, Gilbert both summarized and questioned the many medicinal uses of lodestone and iron during this time. Gilbert focused not on the power of magnetism as a force to cure diseases, but on the description and use of iron in the practice of medicine. In this regard, Gilbert was making an early distinction between forces of nature and matter. As a pioneer in the process of experimentation, Gilbert's writings warned against existing practices of searching for the remedies of diseases without first focusing on, and investigating, the causes of the illness.

Gilbert did not confine his study of magnetism to medicine. Like many of the scientists of the 17th century, Gilbert had training in a variety of scientific disciplines. In *De Magnete*, a publication on magnetic principles released in 1600, Gilbert described the Earth as a spherical magnet with distinct poles that functioned the same as a magnet. Gilbert characterized the changes in the magnetic field, called magnetic declination, as one moved from the equator toward the poles (see MAGNETIC DECLINATION). While he indicated that this change did correspond to changes in latitude, he erroneously thought that the geographic and magnetic poles of the Earth were located at the same location. This error was probably due to a lack of distinction between the attractive forces of magnetism and gravity. The differentiation between these forces would be examined later in the century (see GRAVITATION). Gilbert did go on to describe the magnetic force as a field, and as such would extend beyond the surface of the magnet (see Figure 17). He indicated not only that this force was dependent upon the mass of the object, but that the force of the field would decrease as one moved away from the source.

While Gilbert's concepts on magnetism would not attract the interest of the scientific community until the next century, his portrayal of

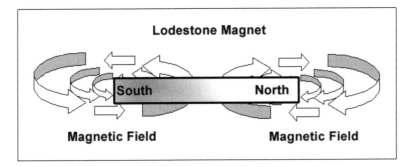

Figure 17. The projection of the magnetic fields from a lodestone magnet, as described by Gilbert (1600). The north and south poles of the magnet were first named by Peregrinus in the 13th century.

magnetism as a measurable force of nature would have considerable influence on the next generation of scientists. The 17th century was a time when scientists sought to define the natural forces of the universe in mathematical terms. In the later half of the century, a few scientists would apply Gilbert's ideas on magnetism to other natural phenomena. For example, around 1641 the German scientist Athanasius Kircher (1602–1680) attempted to explain the tides as a function of the magnetic properties of the Earth and its moon (see OCEAN TIDES). Still others sought to further define the mathematical principles of magnetism, such as the Italian scientist Nicolas Cabeo's (1586–1650) work with the propagation of magnetic forces.

Gilbert's publications, specifically *De Magnete*, had a significant impact on the development and thinking of notable scientists such as Johannes Kepler and Galileo. Galileo was specifically attracted to Gilbert's new use of mathematics to explain the natural world. Kepler would expand on Gilbert's magnetic principles of the Earth to develop his own models of the solar system. In doing so, Kepler would suggest that the elliptical orbits of the planets were the result of magnetic forces between the sun and the planets, and that these forces influenced the velocities of the planets (see KEPLER'S THEORY OF PLANETARY MOTION). In turn, Kepler and Galileo would have a direct influence on the work of Isaac Newton and his theories of universal gravitation (see GRAVITATION).

However, the discovery of magnetism in the 17th century did not have a direct impact on society at large. The mystical properties of magnets would continue to have a place in medicinal practices for a considerable time after Gilbert. It is interesting to note that 400 years later, supporters of alternative medicine claim that magnetic stones have healing properties, despite the lack of experimental evidence supporting the claims. However, while magnetism does not appear to have the power to cure illnesses, its principles have found widespread use in disease diagnosis, specifically in the use of nuclear magnetic resonance imaging (NMRI). In the 17th century, however, magnetism was one of the first of the natural forces to be explained by the new generation of mathematical scientists. Whereas Gilbert's process of scientific investigation, one that relied on experimental data to confirm ideas, was unique at the beginning of the century, it quickly found supporters in the new breed of experimental scientists of the 17th century. By the dawn of the 18th century the experimental approach had developed into a common factor for research in all the major scientific disciplines.

Selected Bibliography

Ronan, Colin. *Science: Its History and Development among the World's Cultures.* New York: Facts on File, 1982.

Spangenburg, R., and D. K. Moser. *The History of Science from the Ancient Greeks to the Scientific Revolution.* New York: Facts on File, 1993.

Taton, Rene, ed. *The Beginnings of Modern Science.* New York: Basic Books, 1958.

Wolf, A. *A History of Science, Technology and Philosophy in the 16th and 17th Centuries.* New York: Macmillan, 1968.

Mars (1609–1672): The planet Mars has occupied a place in mythology since the time of the earliest star watchers. The distinct red color of Mars makes it easily distinguishable in the night sky. Surpassed only by the sun, moon, and Venus in brightness, Mars has historically been associated with death and war. In the Roman Empire, Mars was the name of the god of war, the Romanized version of the Greek god Ares. To the Romans, the god Mars served as the protector of the Roman emperor and was second in importance only to the king of the gods, Jupiter. In 17th-century astronomical science the planet Mars would be second only to discoveries involving the planet Jupiter (see JUPITER). During this century Mars would serve as the testing ground for new theories of planetary motion and early calculations on the dimensions of the solar system.

Prior to the 17th century, the planet Mars had already established a reputation for itself. To the ancient Greeks, Mars represented one of the five known planets. To these early astronomers, Mars appeared to have a mind of its own as it moved across the night sky; thus it was frequently called a wanderer. The planet was not only capable of slowing its movement but on certain occasions it actually moved backward against the background of stars. While this retrograde motion was not unique to Mars, as Jupiter and Saturn also exhibited a similar movement, the effect with Mars was much more pronounced. To the ancient Greeks, the establishment of order in the physical world was a prime concern, and Mars greatly complicated early attempts to construct models of the solar system. The Greek model of a solar system centering and revolving around the Earth, called a geocentric model, could easily explain the motions of the stars, moon, and sun. The motions of Mars, however, could only be made to fit the geocentric model by the addition of complicated correction factors, called epicycles, to its orbit (see KEPLER'S THEORY OF PLANETARY MOTION).

By the 16th century a millennium of accumulated corrections to the orbits of the planets had made the geocentric system almost unusable. In its place new theories were being developed. The simplest of these would be a heliocentric, or sun-centered, system (see HELIOCENTRISM). However, this system did not appropriately place man as the center of the universe, and thus was not popular with the ruling religious groups of Europe. Tycho Brahe (Tycho), a Danish astronomer, attempted to revive the old geocentric system by adding new calculations and astronomical data. For decades Tycho accumulated observations on the motions of the planets, with his attention focused on Mars. His pretelescopic records of Mars are considered some of the most detailed of the time, and provided a historical database on the movement of the planet. Early in the 17th century Johannes Kepler, a student of Tycho's but a strong supporter of heliocentrism, was convinced that Tycho's data provided insight into determining the true nature of planetary orbits. Over a period of several years, Kepler conducted detailed calculations of the Martian orbit. Initially Kepler was convinced that the shape of planetary orbits would be circular, a remnant of the ancient Greek philosophy that the motions of the planets were perfect circles. However, despite his best attempts Kepler could not get the observations to precisely match a circular model. Determined to solve the mystery, Kepler decided to examine the Earth's orbit from the perspective of Mars. He knew that it took Mars 687 days to complete its orbit, while the Earth required only 365.26 days; then at 687-day intervals Mars would return to the same place in its orbit, while the Earth would be at a different location. By recognizing the location of Mars in the night sky on these two dates, it was possible to derive the angle of Mars relative to the sun (a fixed reference point). This gave Kepler an angular measurement to work with, and after conducting a series of formidable calculations, he was able to determine the distances between the sun and the Earth and Mars. It is significant to note that these calculations were done without the aid of calculating machines (see CALCULATING MACHINES), although Kepler may have had access to early logarithms (see LOGARITHMS). Since the distances between Mars and the sun varied with the position of Mars in its orbit, Kepler was forced to abandon the circular orbit in favor of an elliptical-shaped one. In the process of examining Mars, Kepler defined three laws of planetary motion that formed the basis of astronomy until the time of Isaac Newton (see KEPLER'S THEORY OF PLANETARY MOTION).

To many in the 17th century, Earth was considered to have a special place in the heavens. While Kepler had demonstrated that the orbit of Earth followed the same rules as the other planets, the Earth and moon were the only known places in the solar system on which surface features could be distinguished. The invention of the telescope in 1609 would forever change observational astronomy as these new instruments would make it possible to examine the planets in greater detail (see TELESCOPE). Around 1610, Galileo became the first to determine that Mars was not merely a light in the sky, like a star, but that it had a planetary disk. However, since these early telescopes were plagued by focusing and chromatic (color) problems, more detailed examinations of Mars were difficult to obtain. Even Galileo, an experienced astronomical observer, would be fooled by the poor optical systems of these telescopes into believing that Saturn was actually three planets (see SATURN).

By the mid-1600s a number of improvements had been made to the telescope, as well as the invention of specialized instruments to aid observation (see MICROMETER; PENDULUM CLOCK). Perhaps the most important technological advance was in the improvement of lens-grinding techniques. In 1659, using an improved lens that he had invented, Christiaan Huygens turned his attention to Mars. Soon he discovered a smudge on the surface of the planet that remained present in subsequent observations. Huygens named this feature Syrtis Major (large bog) and it became the first physical feature on a planet to be identified outside of the Earth-moon system (see Figure 18). Huygens also noted that the spot appeared about every 24 hours, suggesting that Mars rotated on its axis in about the same period as the Earth. In 1665, Giovanni Cassini calculated the rotation of Mars to be 24

Figure 18. Modern image of the Syrtis Major region of planet Mars obtained by the Viking missions (ca. 1975). (Image courtesy of NASA, the National Space Science Data Center and Dr. Michael Carr.)

hours 40 minutes, very close to the actual value of 24 hours 37 minutes. Several years later, in 1672, Huygens was the first to describe the polar caps on Mars, although detailed observations of the poles would have to wait until studied later in the 18th century. In many regards, the appearance of Mars was beginning to be very similar to that of the Earth, a fact that supported the developing philosophy that our planet was not unique.

While Kepler had used Mars to establish his laws of planetary motion and had done some early calculations on distances, no one had yet begun the task of determining the dimensions of the solar system. In 1672, Cassini, Jean Picard, and Jean Richer set out to determine the distance to Mars using a method called parallax analysis. If an object is viewed from two different locations, it appears to shift its position in relation to the background. Objects that are near shift more when viewed from two vantage points, while distant objects exhibit less of a shift. If the distances between the observation points and the observed angle to the object are known, then it is possible to calculate the distance to the object. These values form the baseline and two vertices of a triangle, and once this is established the remaining values of the triangle may be determined. The Greek astronomer Hipparchus had used this procedure around 190 B.C.E. to determine the distance from the Earth to the moon. In the 17th century, Cassini, Picard, and Richer would use basically the same procedure. Observations of Mars, aided by the use of pendulum clocks, were made from Paris and French Guiana. This information allowed the distance to Mars to be calculated. Using Kepler's laws of planetary motion (see KEPLER'S THEORY OF PLANETARY MOTION), it was then possible to calculate the distance from the Earth to the sun, a distance now called an **astronomical unit** (frequently abbreviated A.U.). Despite primitive equipment, Cassini's calculation of eighty-seven million miles was close to the actual distance of about ninety-three million miles. Of greater importance, it had become obvious that distances in the solar system were massive. Saturn, the most distant object then detected, was over 1,600 million miles away. The stars were so far distant that parallax measurements were impossible. Once the distances to the planets were known, it was then possible to calculate the size of the planets using another invention of the 17th century, the micrometer (see MICROMETER). The immense size of the gas giants and the vast distances of the universe would sever any remaining ideas that the Earth was the center of the universe.

Discoveries of the nature of Mars and its two satellites would continue past the 17th century. Through the following centuries Mars would be the center of controversy and speculation on extraterrestrial life. Three hundred years after Huygens and Cassini began the exploration of Mars, a series of unmanned American and Soviet space probes have supplied evidence that Mars may have closely resembled Earth in its past. It is possible that Mars may have possessed running water and the precursors of life. Modern expeditions to Mars, including manned missions, over the next several decades may provide answers to questions on the evolution of life in our solar system. The fact that Mars remains a focus of scientific discovery and exploration in this century is a direct tribute to the scientific pioneers of the 17th century who first investigated the planet and learned of its similarities to our own.

Selected Bibliography

Evans, James. *The History and Practice of Ancient Astronomy*. New York: Oxford University Press, 1998.

Gingerich, Owen. *The Eye of Heaven: Ptolemy, Copernicus and Kepler*. New York: American Institute of Physics, 1993.

Motz, Lloyd, and Jefferson H. Weaver. *The Story of Astronomy*. New York: Plenum Press, 1995.

Sheehan, William. *The Planet Mars: A History of Observation and Discovery*. Tucson: University of Arizona Press, 1996.

Trefil, James. "Puzzling out Parallax." *Astronomy* 26, no. 9 (1998): 46–51.

Meteorology (1625–1660): Meteorology is the study of atmospheric phenomena, and as a physical science it primarily focuses on those factors that influence the weather. The study of weather predates the time of the 17th century as most ancient cultures had some mechanism for recording excessively hot or cold periods as well as relative amounts of rainfall. Of all the ancient cultures, the Chinese probably were the most advanced, although it is known that the ancient Hindus also kept some records. The Chinese had an intense interest in the relationship between natural phenomena and important cultural events. For this reason the Chinese kept detailed records of both weather-related and astronomical events. For example, the Chinese were well recognized in their documentation of ocean tides and the cycles of asteroid showers and comets (see HALLEY'S COMET; OCEAN TIDES). This level of detail was not repeated in Western cultures until the time of the ancient Greeks. The Greek philosopher Anaximander proposed some early causes of lightning and thunder.

Unfortunately, most early Greek meteorology involved the regulation of weather events by deities, which greatly lessened the influence on later generations of scientists. Some progress was made with Aristotle's *Meteorology*, although it included several topics not now associated with the science of meteorology. Regardless of these early works, accurate quantitative measurement of meteorological events, such as precipitation, would have to wait until the invention of dedicated meteorological instruments in the 17th century.

The 17th century was a time of great innovation in the invention of scientific instruments. Early in the century the first thermometers (1612) were invented by the Italian physicist Sanctorius (see THERMOMETER). By 1644 the work of Evangelista Torricelli had made it possible to detect changes in barometric pressure using a barometer (see BAROMETER). Later in the century (ca. 1657) the invention of the hygrometer made it possible to measure the relative humidity of the atmosphere (see HYGROMETERS AND HYGROSCOPES). In addition to the invention of these devices, there were several other meteorological instruments whose modern origins may be traced to the 17th century. Although rain gauges had been used in the Far East since the time of the 15th century, the Western invention of a rain gauge in 1662 by the English architect and inventor Christopher Wren introduced the instrument to modern studies of weather. Improvements in the accuracy of the instrument were made by a number of investigators, including an instrument designed in 1695 by the English inventor Robert Hooke. This rain gauge lessened the influence of evaporation on the measurement by incorporating a funnel that drained the precipitation into a narrow-necked flask.

Another instrument that saw refinement during the 17th century was the anemometer, or wind gauge. Most wind gauges are relatively simple devices in which the pressure of the wind moves a pointer to indicate wind direction. Other anemometers determine wind velocity by utilizing the force of the wind to turn a windmill-like device. Although Leonardo da Vinci is often credited with the invention of an anemometer in the 15th century, several 17th-century scientists also contributed to its design. Sanctorius developed an anemometer in which the force of the wind was directed against a vertical plate. Although mechanical modifications to its design were made over the next two centuries, the basic principles served as a basis for instruments into the 20th century. Finally, the invention of the pendulum clock in 1656 by Christiaan Huygens provided scientists with the ability to make more precise measurements of time (see PENDULUM CLOCK). A version of the pendulum clock was adapted for the study of weather in 1664 by Hooke. This clock not only kept accurate time, but also measured precipitation, barometric pressure, wind velocity and direction, and relative humidity. It also possessed the ability to make a record of the measurements, although with what level of accuracy is unknown, using a rotating drum and paper punches.

The science of meteorology in the 17th century was not limited to the construction of scientific instruments. Several attempts were made during this century to apply these instruments in an organized fashion to the study of weather. Periodic weather observations had been attempted in the 16th century in Germany, but on a rather small scale. Many scientists of the 17th century at least experimented with weather analysis using the new generation of scientific instruments. Kepler, Galileo, and Descartes all made weather observations at some point in their careers. By 1660 both Robert Hooke and the French scientist Blaise Pascal are known to have observed the changes in barometric pressure caused by an approaching storm. Robert Hooke also contributed by making up one of the first weather charts, which he called the Form of a Scheme. This attempted to make a connection between observed weather patterns and natural phenomena such as the wind directions and phases of the moon. In the mid-1600s the Academia del Cimento, an Italian scientific society under the direction of the Grand Duke Ferdinand, was commissioned to conduct weather observations at various locations in Europe. Initially established to monitor the weather in seven cities, it was expanded to eleven cities within a few years of its inception. For almost a decade the observers recorded temperature, barometric pressure, wind conditions, and relative humidity and forwarded this information to a central location for analysis. However, this venture was short-lived and would not be repeated on a large scale again until the late 1700s. Long-term weather observations from a single location were initiated by the French Academy of Sciences around 1666 in Paris. This was to be the longest-duration series of weather observations for the times, with consistent weather readings being taken by a member of the academy well into the 18th century.

The study of the atmosphere was a prime area of research during the 17th century with a large number of scientists making contributions to the field. Studies of atmospheric pressure and the nature of gases, as well as the invention of the barometer, all demonstrated an interest in understanding the natural forces of the planet (see BAROMETER; BOYLE'S LAW; GASES). The application of this work to the study of weather was a logical progression for the science of the times. Although modern meteorological practices and instruments, including satellite imaging and computer-generated models, bear little resemblance to the science that originated in the 17th century, many of the instruments that are in use today follow the same principles as those invented over three centuries ago. However, weather remains a chaotic force of nature and thus is frequently difficult to predict simultaneously at the local, regional, and national levels. It is likely that the 21st century will continue to develop innovative scientific methods of analyzing the weather in much the same manner as was begun in the 17th century.

Selected Bibliography

Bud, Robert, and Deborah Jean Warner, eds. *Instruments of Science*. New York: Garland Publishing, 1998.

Frisinger, Howard H. *The History of Meteorology: to 1800*. New York: Science History Publications, 1977.

Taton, Rene, ed. *The Beginnings of Modern Science*. New York: Basic Books, 1958.

Wolf, A. *A History of Science, Technology and Philosophy in the 16th and 17th Centuries*. New York: Macmillan, 1968.

Micrometer (1638–ca. 1660): The astronomers of early civilizations noted that the arrangement of stars in the night sky followed predictable seasonal patterns. Thus, with careful observation, they could be used as a type of calendar to signal important annual events. Early in the history of mankind the appearance of a specific astronomical pattern might have indicated the time for migration in search of new food sources. After the development of agricultural practices it became possible to synchronize the planting and harvesting of crops with the seasonal changes heralded by the stars. These patterns, called constellations, played an important part in the success of many early cultures. The ancient Egyptians may have used the appearance of stars and constellations, specifically the star Sirius in the modern constellation Canis Major, as a predictor of the annual flooding of the Nile. Because of the impact that the constellations had on the success or failure of a civilization there often developed a belief that the constellations were manifestations of the gods. Thus considerable effort was put forth by early astronomers to map and predict the occurrence of a constellation.

Charts of the constellations became some of the first examples of the making of star maps (see STAR ATLASES). While the current star maps are based on the names of the constellations as provided by the ancient Greeks, there is evidence to suggest that the names and myths associated with Greek origin actually originated in the Middle East and were transferred to the Greeks via the Phoenicians. Evidence of star maps may be found in the early Mesopotamian and Egyptian civilizations. The Chinese had also developed star atlases, some of which were fairly advanced by the 10th century. However, while these maps often accurately displayed the configuration of the stars, there did not exist a consistent method of measuring the distances between the stars against the backdrop of the night sky.

The invention of the telescope early in the 17th century had a significant impact on astronomers (see TELESCOPE). With the aid of telescopes, astronomers began to probe the complexities of the solar system, an ability that quickly resulted in the demise of long-standing models of how the solar system is organized. But more importantly, with the design of the telescope a need arose for additional scientific instruments to assist with the recording of astronomical data. Specifically the early users of the telescope required two instruments with which to advance the accuracy of their observations, a method of recording accurate time measurements and a mechanism for measuring astronomical distances.

The first of these problems, that of time, had long been a concern of astronomers. Many of the earliest timekeeping devices, be they water or weight driven, were initially designed for astronomers. However, the invention of a functional pendulum clock in 1656 by Christiaan Huygens greatly influenced astronomical work. The pendulum clock accurately measured fractions of an hour, down to minutes, that were not possible with earlier timepieces. Since the clock was based on gravitation, which for the most part is constant at sea level, the resulting measurements would be consistent between instruments (see PENDULUM CLOCK). With the pendulum clock it was now possible to more precisely calculate the duration of astronomical observations, such as the period of rotation for a planet or moon.

The second problem, that of distance, would require more innovation. It was possible to calculate the distance from Earth to the planets by a process called parallax analysis. The basic premise of parallax analysis is that objects that are close to the viewer appear to shift in relation to the background when viewed from different vantage points. However, objects that are more distant appear to remain stationary when viewed from different angles. It had been noted that the parallax effect was minimal when applied to the stars, suggesting that these objects were a considerable distance from the Earth. So the distance from the Earth could not be used as a means of generating accurate maps, since that distance was not calculable for stars. What was needed was a standard for determining the distance between two stars against the background of the night sky. Prior to the **micrometer** star maps were constructed using measurements determined by the naked eye (see STAR ATLASES). While some of these maps possessed high levels of detail, such as those constructed by Tycho Brahe and Johannes Hevelius in the late-16th and early-17th centuries, the improvements in telescope design and magnification required new measurement instruments, which had to be compatible with the construction of the early telescopes as well as provide measurement that could accurately be compared between observers.

The first micrometer was a relatively simple instrument. A small device was constructed using two pieces of thin metal on the edges. The device was then placed inside the telescope at the focal point of the incoming light. By turning a small metal screw, the thin pieces of metal that formed the edges of the micrometer could be moved inward or outward as needed. The screw itself was calibrated so that the change in distance caused by one turn of the screw was known. To use the micrometer, one edge was placed on the starting object, and the screw adjusted so that the other edge resided on the second object. The distance bracketed by the two edges of the micrometer, as determined by the calibrated screw, could then be used to calculate the angular distance between the objects given that the magnification power of the telescope was known. This simple device was first constructed by an English astronomer named William Gascoigne (1612–1644) around 1638. Unfortunately, Gascoigne's work was not widely published and thus remained relatively unknown until later in the century.

The changes in the telescope during the 1650s and 1660s (see TELESCOPE) spurred interest in the development of measuring instruments specific for the telescope. The widespread use of the telescope meant that a number of individuals were working independently on developing a measurement device. By the 1660s a number of individuals had duplicated Gascoigne's work, including Christiaan Huygens, Adrien Auzout, and Robert Hooke. The later pair of astronomers used fine hairs in place of the thin metal pieces used in the Gascoigne device. The invention of a similar instrument simultaneously by multiple scientists is a significant indicator of the necessity for such a device. The invention of the micrometer was not a scientific breakthrough in itself, but rather a technological enhancement of an existing instrument. However, it was an indicator of the need for a higher degree of accuracy in astronomical measurements.

Once invented, the micrometer made it possible for more precise astronomical calculations. The French astronomer Jean Picard may have been one of the first to use the micrometer when in 1667 he made his calculations on the diameter of the Earth. This calculation would have an important impact on the verification of Isaac Newton's theories on gravity (see GRAVITATION). Later in the century Giovanni Cassini used the pendulum clock and the parallax effect to gain some insight on the dimensions of the solar system. By measuring transit times of the planets from different observational points on Earth, Cassini and others were able to determine the distances of the planets from the Earth. Once this was established, the micrometer could then be used to determine the planet's diameter. In doing so, the relative size of Earth in comparison to the other known planets could be determined. The establishment of these special relationships would become important in making some initial calculations on the speed of light. In 1675 Olaus Roemer, using Cassini's data on the width of the Earth's orbit and his own calculations on how long it took light to pass through Earth's orbit, determined the speed of light to be in excess of 140,000 miles per second (227,000 kilometers). While not completely accurate (the actual speed is slightly over 186,000 miles per second), it was an important first step in measuring the speed of light (see LIGHT SPEED).

Breakthrough scientific achievements often foster the development of new classes of instruments designed to improve upon the initial discovery. The invention of the micrometer in the 17th century is an excellent example. The invention of the telescope early in the century, and the subsequent improvements in its design, gave rise to the development of instruments such as the micrometer and the pendulum clock. Together, these instruments gave astronomers the ability to make some initial calculations on the physical properties of the solar system. While these calculations often were not entirely accurate, they did provide an important first step in quantifying astronomical observations and would set the stage for more precise measurements in the next century. Use of micrometers would not be limited to astronomy. Micrometers are now used by microbiologists to measure the size of microscopic organisms and to improve the accuracy of gun sights. They have also found application in industrial work where they can be used to measure fine distances in milling work where precise engineering specifications must be met.

Selected Bibliography

Asimov, Isaac. *Eyes on the Universe: A History of the Telescope.* Boston: Houghton Mifflin, 1975.

Bud, Robert, and Deborah Jean Warner, eds. *Instruments of Science.* New York: Garland Publishing, 1998.

Wolf, A. *A History of Science, Technology and Philosophy in the 16th and 17th Centuries.* New York: Macmillan, 1968.

Microscope (ca. 1600–ca. 1683): The rise of the scientific revolution, and the increased use of mathematics to explain the laws of the natural world, led to a demand for more sophisticated scientific instruments. The evolution of magnification devices in the 1600s was a significant development for scientific research. The 17th century saw two distinct phases in the development of microscopes. In the opening decades, microscopes were being designed and perfected from crude optical devices such as spectacles and the early telescopes. However, by the second half of the century, microscopes were being used to investigate a new realm in biology, that of cells and microorganisms.

The magnification ability of substances such as water is well documented in the history of science. As early as the 2nd century B.C.E., the Greek philosopher Ptolemy had described the ability of water to bend light and provide some magnification of objects. Ptolemy empirically measured the ability of water to bend light, a rare event for a time when hypothetical thinking took dominance over experimentation. About the same time, there were indications that glass could be used to concentrate light for the purpose of making fires and possibly even the cauterization of wounds. However, it would not be until centuries later when the Arabic scholar Ibn al-Haytham, also known as Alhazen, would be one of the first to describe the magnification abilities of a lens. This began what was to later become the study of optics, or the study of the characteristics and properties of light (see OPTICS). In the 13th century, Roger Bacon (1214–1294), an English philosopher and alchemist, applied the principles set forth by Alhazen to design the first optical instruments for the correction of vision. These were called spectacles. Unfortunately, these primitive devices used a single lens and typically did not provide a significant amount of magnification.

The first true microscopes, those with the ability to resolve microscopic organisms, were most likely derived from the recent inventions of the telescope in Europe. While there is considerable controversy over the origins of the first microscope, it is known that in the late 1590s, Hans Lippershey (1570–1619) and Zacharias Janssen (1588–1630), Dutch eyeglass makers, used a two-mirror system to design the first functional telescope for viewing objects at a long distance (see TELESCOPE). While credit for converting the principles of a Janssen telescope into a microscope is often bestowed upon Galileo, it is unlikely that he was the original inventor. Rather, it appears that Galileo made improvements on the optical systems of existing telescopes and microscopes. However, the term *microscope* appears to have been coined from an academic society of which Galileo was a member. Regardless of the origin, the importance of the microscope for scientific observations was immediately noticed throughout Europe, and in just a few decades there were numerous microscope manufacturers on the continent.

The microscopes of the early 1600s were crude magnification devices. The microscopes of this time either were simple, using a single lens, or utilized a compound optical system. A compound device utilizes two, or more, lenses to increase the resolution of the image. While these early instruments were improvements over the magnification potentials of single lenses such as those used in spectacles, they often consisted of imperfectly ground lenses and cardboard construction. These instruments typically had a magnification potential of only three to nine times normal viewing. These primitive instruments had little resemblance to the microscopes found in science labs today. In fact, some were little more than handheld magnifying glasses, while others stood over two feet tall. Over the next several decades, improvements were made to both the optical systems and mounting hardware of the instruments. However, it is not until 1625 that Francesco Stelluti (1577–1640), an Italian scientist with an interest in insect anatomy, made the first known publication of data derived from microscopic investigations.

The following decades saw the formation of a new discipline of science focused on microscopic observations. One of the first of this new group was the Italian microscopist Marcello Malpighi. During the time of Malpighi there was considerable controversy and investigation on the anatomy and physiology of the circulating system (see BLOOD CIRCULATION). The prevailing idea of the time was that blood flow was unidirectional. Blood was carried by both veins and arteries directly to the tissues of the body, where it was then consumed. Others of this time, such as the English physician William Harvey, contended that the veins and arteries formed a closed circulatory system in which the blood circulated. What was missing in Harvey's model was identification of the connecting vessel between the veins and arteries of the body. Intrigued by Harvey's work, Malpighi focused on finding the mechanism by which blood was transferred between the major vessels of the circulatory system. Around 1661, using a compound microscope to examine the thin membranes of bat wings, Malpighi discovered small vessels that connected the veins and arteries within the tissues. The structures that Malpighi observed were later called capillaries and their discovery not only firmly proved Harvey's concept of a closed circulatory system, but also confirmed the use of microscopes for scientific observations.

Around the same time as Malpighi, the English physicist Robert Hooke was turning his attention to the microscopic world. Hooke made several important contributions to both the development of the compound microscope and the emerging science of microscopy. As a physicist, Hooke was interested in not only the optical properties of a microscope, but also the physical construction of such devices. Hooke made improvements in illumination systems, mechanical construction of the body of the microscope, and the invention of a mechanical stage to hold the specimen in place during observations. He was one of the first microscopists to attempt to estimate the size of objects under a microscope. However, Hooke's greatest contribution to the use of the microscope as a scientific instrument came in 1665 when, under direction of the Royal Society, he published *Micrographia*. This publication introduced the word "cell" to the scientific community (see CELLS). Hooke used the term to describe the honeycomb-like structure of the cork plant when observed under a microscope. *Micrographia*, with its superior illustrations of microscopic observations, probably furthered the causes of microscopy more than any other publication of its time.

While Malpighi and Hooke were undoubtedly the pioneers in the use of the microscope for scientific observations, Anton van Leeuwenhoek was the individual who perfected the science of microscopy. Leeuwenhoek, a Dutch merchant by trade and an amateur scientist by desire, not only improved on the design of the microscope, but also made several significant scientific contributions to the field. While Leeuwenhoek worked with simple rather than compound microscopes, the microscopes he designed were far better in quality, with magnifications of almost 200 times normal view. The craftsmanship of the Leeuwenhoek microscopes, sometimes called "flea glasses," introduced a new era in microscope construction. From this time forward, microscopes would be made as finely prepared instruments. The high level of craftsmanship in the Leeuwenhoek devices allowed for highly detailed microscopic examinations, and thus revealed a hidden world of cells. Without the scientific background of most scientists of his time, Leeuwenhoek did not attempt to explain his observations, but rather focused on making detailed descriptions of what he saw. Leeuwenhoek completed Malpighi's earlier observations by describing the flow of blood within a capillary. Among his many discoveries were the first observations of bacteria, protozoa, and sperm cells.

The development of the microscope in the 17th century would have a significant impact on all aspects of the sciences, but particularly the life sciences. The microscopes of this century were some of the first of a new generation of scientific instruments designed for specific tasks. While the next generation of scientists would improve on the design and function of the microscope, the 17th century marked the evolution of the instrument into an effective observation tool. Furthermore, the development of the microscope would result in the formation of a new group of scientific investigators, the microbiologists. Modern microscopes are vastly more powerful than their predecessors, but still function according to the same basic magnification principles. Exceptions are the transmission and scanning electron microscopes, which utilize electrons in place of light. However, the goal of these devices in viewing the microscopic world has not changed since the time of the 17th century.

Selected Bibliography

Bradbury, S. *The Evolution of the Microscope.* New York: Pergamon Press, 1967.

Bud, Robert, and Deborah Jean Warner, eds. *Instruments of Science.* New York: Garland Publishing, 1998.

Hall, A. Rupert. *From Galileo to Newton.* New York: Dover Publications, 1981.

Spangenburg, R., and D. K. Moser. *The History of Science from the Ancient Greeks to the Scientific Revolution.* New York: Facts on File, 1993.

The Moon (1609–1687): The early decades of the 17th century were an important time for the study of astronomy and physics. Scientists during this time were questioning the structure of the universe as presented by the ancient Greeks centuries earlier. In this period of transition the moon played a crucial role as the testing ground for new ideas regarding the structure and nature of the solar system. The philosophy of the ancient Greeks held that the heavens represented perfection, with all that is imperfect and bad confined to the Earth. The moon and stars were perfect spheres and were believed to orbit the Earth in perfect circles (see KEPLER'S THEORY OF PLANETARY MOTION). However, the close proximity of the Earth to the moon was believed to negatively distort the perfect nature of the moon's surface, resulting in blemishes that could be viewed with the naked eye. In addition, Greek mathematicians, such as Aristarchus and Hipparchus, had utilized their advances in geometry to calculate the distance to the moon with a surprising degree of accuracy. The Greek model of the universe persisted for almost 2,000 years, despite the accumulation of evidence suggesting that it was in error (see HE-LIOCENTRISM; JUPITER). Challenges to most of these ideas began with the Renaissance and its renewed interest in science and mathematics. However, it was the invention of the telescope (see TELESCOPE) in the 17th century that initiated a wave of scientific investigation focused on examining the true nature of the solar system.

One of the first to turn a telescope toward the moon was the Italian scientist Galileo. Galileo's initial view of the moon through the telescope yielded an amazing amount of detail. The moon clearly contained both highlands and lowlands. These are frequently called seas, even though the moon has never possessed a significant amount of water. This amount of detail provided the first evidence that the moon was a body similar to the Earth, and not an object of divine perfection. Inspired by this discovery, Galileo turned his attention to the study of Jupiter. This resulted in the discovery of the Galilean moons and Galileo's support of a sun-centered solar system (see JUPITER). Galileo made some initial maps of the moon and even tried to estimate the size of some of the mountain peaks by calculating the length of the shadows in relation to the angle of the sun. However, the primitive nature of these early telescopes greatly reduced the accuracy of his work. As the 17th century progressed, improvements in the optical design of the telescope reduced some of the focusing problems. The invention of the long telescope by Johannes Kepler enabled astronomers to view the night sky with greater clarity (see TELESCOPE). In 1647 the Polish/German astronomer Johannes Hevelius used a long telescope to make an early map of the moon. His map named many of the mountain ranges and the major seas of the moon, for example the sea Mare Serenitatus. Hevelius also named a number of craters. However, in 1651 the Italian astronomer Giovanni Battista Riccioli (1598–1671) published a new protocol for naming craters that honored past scientists and astronomers. It is this system that remains in use today. It had long been known that the moon always presents the same face to the Earth, owing to the time synchronization of its orbit around the Earth and rotation on its axis. However, Hevelius discovered that the moon was not always stable in its rotation; thus it was possible at times to view a little into other hemispheres. These instabilities are frequently called the **librations**, or wobble, of the moon. Because of this, slightly more than 59 percent of the moon's surface is visible from Earth.

The nature of the Earth-moon system, specifically those forces that are responsible for distance between the two bodies, was a focus of 17th-century physical scientists. The scientific description of magnetism as a force of nature in the early 17th century by William Gilbert caused a large number of scientists to believe that it was this force that was responsible for the structure of the solar system (see MAGNETISM). One of these was Kepler, who was investigating the motion of the planets. Kepler recognized that the motion of the planets was due to the interaction of two attractive forces. In the case of the Earth-moon system this force was projected outward from both the Earth and moon. While he recognized that this force decreased with distance, he erroneously thought the force to be magnetic in nature (see KEPLER'S THEORY OF PLANETARY MOTION). The German-Italian astronomer Athanasius Kircher further expanded the magnetic power of the moon when he proposed that ocean tides were the result of a magnetic attraction by the moon (see OCEAN TIDES). While Kircher had the principle correct, he had also misidentified the force. However, later in the 17th century Newton's work on gravitation would provide the link between the moon and tides (see GRAVITATION).

Perhaps the greatest contribution that the moon made to 17th-century science was as a proving ground for Isaac Newton's developing ideas on the laws of motion and gravitation. Earlier in the century Galileo had discussed the principles of inertia, velocity, and acceleration as they related to the physics of a falling object (see LAW OF FALLING BODIES). His mathematical derivation of this law influenced Newton in his later work on his laws of motion. Newton derived three basic laws of motion, the first of which was fundamentally a restating of Galileo's principles of inertia. Using these laws, Newton proposed that the unifying force responsible for the motion of the planets and the moon was gravitation (see GRAVITATION). But Newton did not simply just speculate on the nature of this force, he used the Earth-moon system as a confirmation of his synthesis of motion and gravitation. Newton recognized that according to the principles of inertia the motion of the moon should be in a straight line. Thus a second force must be bending the motion of the moon into its current orbit. Newton suggested that this force was gravity and that the force of gravity decreased with the distance from its source according to the inverse-square law (see GRAVITATION). By measuring the path of the moon and the rates of acceleration on both the Earth and moon, Newton was able to mathematically confirm the contribution of gravitation as the governing force. Newton then correctly assumed that this force was the unifying force behind the structure of the entire universe.

Throughout recorded history the moon has held an important place for mankind. As an early timepiece to record the passing of seasons to the view of the moon as a representative of the perfect nature of the heavens, the moon has always been a focus of scientific investigation. However, in the 17th century scientists brought new technological advances, such as the telescope, and innovative mathematical procedures to study the physical forces of nature. The moon played a key role in this process and continues to do so until this day. In our time the moon has been extensively probed and explored, and used to understand the age and violent history of our solar system. The visits by manned spacecraft in the 20th century once again demonstrated the use of the moon as a testing ground for new technology and scientific achievements, a role that it has held since the 17th century.

Selected Bibliography

Asimov, Isaac. *Eyes on the Universe: A History of the Telescope.* Boston: Houghton Mifflin, 1975.

Evans, James. *The History and Practice of Ancient Astronomy.* New York: Oxford University Press, 1998.

Moore, Patrick. *The Great Astronomical Revolution: 1543–1687 and the Space Age Epilogue.* Concord, MA: Albion Publishing, 1994.

Motz, Lloyd, and Jefferson H. Weaver. *The Story of Astronomy.* New York: Plenum Press, 1995.

Wilson, Robert. *Astronomy through the Ages: The Story of the Human Attempt to Understand the Universe.* Princeton, NJ: Princeton University Press, 1997.

N

Nebula (1609–1656): Even with the unaided eye the night sky appears to be full of objects. A casual examination reveals not only an abundance of stars and the predictable motions of the moon and planets, but also the possibility of an occasional appearance by a meteorite or comet. However, with a more detailed look one can detect smudges in the night sky, areas where the stars appear to be blurred. These areas are called *nebulae* after the Latin word for cloud or mist. The early Greek astronomers provide some of the first written descriptions of nebulae in their early preparation of sky charts. The Greek astronomer Hipparchus produced some of the most detailed and significant records of astronomical observations during ancient times. Hipparchus made many contributions to astronomy including assigning brightness, or magnitude, to stars, identifying new stars or novas, and refining calculations on lunar motion and **equinoxes**. This database, which formed the basis of subsequent efforts in Greek astronomy, mentions the positions of two nebulae. Two centuries later, Ptolemy's *Almagest*, a compilation of Greek astronomical and mathematical accomplishments, contains two volumes dedicated to star maps (see STAR ATLASES). In this work Ptolemy lists information on over 1,000 stars, including three additional nebula-like objects. However, further investigation into the nature of these objects would have to wait until centuries later.

The invention of the telescope revolutionized observational astronomy. With the aid of the telescope the vast, complex nature of the heavens was revealed. To the unaided eye, nebulae all appear to be faint glowing clouds or smudges of light. However, with the increased magnification available using the telescope, many nebulae were observed to consist of small groups of stars. In modern astronomy there is a distinction between true nebulae, clouds of interstellar gas and dust, and star clusters. However, in the 17ᵗʰ century there existed considerable controversy on the composition of nebulae. When, in 1609, Galileo observed the first nebula to be detected through a telescope, he disagreed with the notion that they were simply glowing clouds in the heavens. Galileo proposed that all nebulae were simply small groups of stars clustered closely together. To Edmund Halley, however, the nebulae that he observed all possessed an internal **luminescence** that was not due to the action of stars. In actuality, both Galileo and Halley were partly correct, although the answer would take several centuries to resolve. The improvements in astronomical instrumentation in the 19ᵗʰ century would finally allow astronomers to distinguish between Galileo's star clusters and Halley's glowing clouds. Using a **spectroscope**, a device that breaks light into discrete wavelengths for analysis, astronomers determined that true nebulae typically emit a single wavelength of light. In comparison, star clusters emit significantly more complicated light patterns.

The telescopes of the early 17ᵗʰ century were relatively crude instruments, and as such had problems producing quality images. Among the problems were focusing difficulties, called **spherical aberrations**, and problems with colored halos surrounding the images, called **chromatic aberrations** (see TELESCOPE). Despite these inadequacies, the benefits of increased magnification still made new astronomical discoveries possible. In 1612, Simon Marius, a German astronomer, identified the Andromeda Nebula, named because of its location in the constellation Andromeda. In 1924, the American astronomer Edwin Powell Hubble (1889–1953) would determine that this nebula was in fact a large galaxy located over 750,000 light years away. However, as telescope technology progressed in the 17ᵗʰ century, so did image resolution. With the improvements came the ability to detect fainter nebulae. In 1656, Christiaan Huygens, a Dutch astronomer, determined that the smudge of light near the belt of the constellation Orion was indeed a nebula. Now called the Orion Nebula, it is one of the easiest nebulae to distinguish with the naked eye.

While the scientific developments of the 1600s would foster an interest in nebulae, it would not be until the next century that a catalogue of these items would be compiled. The majority of the work in the 18ᵗʰ century would be due to the efforts of two individuals. Charles Messier (1730–1817), a French astronomer, identified over 100 nebula-like objects. Messier had an interest in comets, not nebulae. However, during his time it was possible to initially confuse comets and nebulae. To counter this, Messier developed a catalogue of nebulae, each of which was assigned an identifying number. Many of these numbers remain in use in our times. The second major contributor was William Herschel (1738–1822), who during a thirty-year span recorded the positions of over 2,500 nebulae. The 18th century would also see the origination of the nebular hypothesis. First proposed by the French astronomer Pierre Simon Laplace (1749–1827), this hypothesis suggested that the solar system originated from a rotating nebula. While a revolutionary idea for the 18th century, its basic premises do resemble many modern models of solar system genesis.

Astronomers in our time are continuing the study of nebulae that began during the time of Galileo. Using increasingly sophisticated instruments, including some that are in an Earth orbit, astronomers and physicists are now able to distinguish between different nebular types. Halley's concept of a nebula is now called emission nebulae. In this type of nebula the interstellar gas glows from the ionization of hydrogen gas. An example would be the Orion Nebula, or, using the Messier designation, M42. Galileo's ideas on nebula structure are now classified as reflection nebulae. In a reflection nebula the light of nearby stars is reflected by interstellar dust particles. The Pleiades star cluster, also called the Seven Sisters, is an example of this class. Still other nebulae are considered to be the remnants of a supernova explosion, such as the Crab Nebula (M1). This nebula is the remains of a supernova first detected by the Chinese in 1054. Other classes of nebulae include dark and planetary nebulae. Some of these nebulae are massive in size. For example, the Tarantula Nebula is 1,000 light years across and is actually located in a nearby galaxy. Astronomers now consider nebulae to be the birthplace of stars, and as such allow scientists to gather information on the early history of our solar system. The 17th century played an important role in nebular history as scientists such as Galileo and Halley first questioned their origin. These questions persist to this day and form the basis of present and future astronomical endeavors.

Selected Bibliography

Evans, James. *The History and Practice of Ancient Astronomy.* New York: Oxford University Press, 1998.

Moore, Patrick. *The Great Astronomical Revolution: 1543–1687 and the Space Age Epilogue.* Concord, MA: Albion Publishing, 1994.

Motz, Lloyd, and Jefferson H. Weaver. *The Story of Astronomy.* New York: Plenum Press, 1995.

Ridpath, Ian. *Norton's Star Atlas and Reference Handbook,* 19th ed. Essex, UK: Addison Wesley Longman, 1998.

Wilson, Robert. *Astronomy through the Ages: The Story of the Human Attempt to Understand the Universe.* Princeton, NJ: Princeton University Press, 1997.

O

Ocean Tides (ca. 1610–ca. 1687): Many of the great ancient civilizations originated near one of the oceans or broad seas of our planet. There can be little doubt that the close proximity of these expansive bodies of water had an impact on the development of scientific thought in these cultures. Of the greatest concern to early coastal societies, with regard to the ocean, would have been the ability to predict the occurrence of the daily tides. Navigation, and the subsequent trade routes that the waterways provided, were important to the success of a civilization. The study of tides for early scientists would have been a complex matter, as the timing and magnitude of the tides vary with each coastal location. Early philosophers, most likely those involved with the study of astrology, would have noticed that the position of the moon held some connection with the level of the tides. Since our moon was considered to be a perfect manifestation of the creator, tides were often viewed as the interaction of the gods with the physical Earth (see THE MOON).

While many early cultures developed theories on the origin of tides, the ancient Chinese were among the first to provide written records of their ideas. To the early Chinese civilizations, the tides represented a battle between opposing forces of nature. This was described as early as 300 B.C.E. in *I Ching* (*Book of Changes*) as a battle between the opposing natural forces called yin and yang. By the 2nd century B.C.E., the Chinese had related the tides to the phases of the moon, and by the 4th century they had developed an understanding that the moon was a factor in the occurrence of tides. While around the same period of time the ancient Greeks had also related the phases of the moon to ocean tides, the direct relationship between the two would not be fully established in Western cultures until the time of the Renaissance. The Chinese, however, would continue to lead Western cultures in this area for centuries. For example, the Chinese proposed that the sun contributed to the level of tides by the 11th century, hundreds of years prior to their Western counterparts.

The Dark Ages following the fall of the Roman Empire stifled the majority of scientific thought in Western culture, although a few developments were still being made in the study of tides. In the 7th century the Venerable Bede (ca. 673–735), an English theologian with an interest in timekeeping and the calculation of calendars, noted that not only the phases of the moon but also local conditions had an influence on the timing and height of a tide. This phenomenon, now called the *establishment of the port*, still pertains to local variations in tide tables and levels. While the force that caused tides remained elusive, it was becoming possible to predict tides and construct tide tables. As early as 1213 the English had produced tide tables for the area around the London Bridge.

By the late-16th and early-17th centuries a number of individuals were beginning to focus on the prediction of tides based upon the position of the sun and moon. The prevailing idea when entering the 17th century was that the sun and the moon projected equal attractive forces upon the world's oceans. The force of the sun was relatively constant, affected only by the seasonal changes of the sun's position in the sky. However, the force of the moon was determined by its phase. This attractive force was maximized during the full and new moons and minimized during the times of a half-moon. Thus, the tides would be highest during a full or new moon when the forces of the sun and moon were the greatest, and minimized during the time of the half-moon when the force of the moon canceled the force of the sun. While these ideas were philosophically sound, the predictions did not match the available data on tide levels. The primary problem was in the belief that the sun and moon contributed evenly to the formation of tides and that the forces projected from these two objects were equal. While modern theories acknowledge the contribution of the moon, and to a lesser extent the sun, there are other factors that are involved in the formation of a tide.

A second theory for tide generation focused on the revolution of the Earth on its axis. Historically, Galileo is often considered the first to have proposed this idea. While in fact Galileo did believe that the Earth's rotation was the sole contributing factor in the formation of tides, and that the contributions of the sun and moon were minimal, the idea did not originate with him. There are indications that this theory was considered by philosophers as far back as the time of the Seleuid Empire (ca. 300–100 B.C.E.). However, it was not until the 17th century that the idea gained substantial support. Galileo proposed that the revolution of the Earth on its axis, when coupled to its orbital path around the sun, created oscillations on the floor of the ocean. These oscillations were then transferred to the waters of the oceans and eventually became the tides of the coastal regions. The main problem with this was the monthly variations in the tides. Galileo's model could not account for these variations since both of his contributing factors, the speed of revolution and speed of the orbit, did not significantly vary over time. Thus, this model could not account for the monthly variations in the tides. However, Galileo had added the movement of the Earth as a contributing factor. This concept was further developed by John Wallis, an English mathematician, who later in the century developed theories on the conservation of momentum (see LAWS OF MOTION). Wallis supported Galileo's position by proposing that the monthly tides were an artifact of the interaction of the moon and the Earth. Wallis suggested that the Earth-moon system was not centered on the Earth but rather at a point in space between the two bodies. To Wallis the movement of the Earth and moon around this point as the two circled the sun was responsible for the monthly tides.

While the fact that the moon had an influence on the generation of tides was becoming more accepted by the 17th century, efforts to conclusively prove how the moon interacted with the oceans were being hindered by an inability to identify the physical force of nature responsible for tides. Some believed that tides were caused by the attraction of ocean water to a dry moon, or the interaction of sunlight and moonlight with water. Around the time of the Renaissance, interest was being generated in the study of magnetism, specifically the properties of lodestone and iron ore. Some at the time proposed that the lodestone may be responsible for attracting water toward the moon. However, it was not until the beginning of the 17th century that William Gilbert would experimentally describe magnetism as a measurable force of nature (see MAGNETISM). Many considered magnetism to be the dominant physical force of the universe. In 1654, Athanasius Kircher suggested that the magnetism of the moon was responsible for tides. While Kircher had the force incorrect, he was accurate in predicting that the moon interacted with the oceans via a physical force of nature.

It was not until the science of Isaac Newton that this force would be identified and the relationship between the moon and tides established. In *Principia*, Newton applied his laws of universal gravitation to the persistent problem of tides. The moon, according to Newton, would have the greatest influence on the ocean tides due to the proximity of its mass to the Earth. The gravitational pull of the moon would cause great waves of water to circle the earth once every twenty-five hours, the equivalent of a lunar day. The fact that the tides were not always equal, which would be the case with only a single force, was due to the gravitational influence of the sun. The sun's gravity produces similar waves of water, although they are smaller in magnitude. These waves circle the globe once every solar day, or twenty-four hours. The periodic summation of these waves, not the canceling effects suggested by earlier models, produces the observed variation in tide height. While today it is recognized that other factors are involved in the generation of tides, namely, inertia (see LAWS OF MOTION) and the geographic characteristics of the coastal areas, Newton's explanation had provided a workable solution to the problem.

The study of ocean tides in the 17th century had a unique influence on the science of the time. Before the beginning of the century it was already possible to predict tides based on historical observational data for a coastal area. The determination of the physical force responsible for the cause of tides did not have a major impact on the scientific community in the 17th century. However, the modern maritime industry, with its reliance on supertankers and super-sized container ships, increasingly is dependent on accurate predictions of the tides to move these large ships into and out of ports. In modern science, tides have been recognized to occur on a number of worlds other than Earth. Jupiter's moons Io and Europa (see JUPITER) may exhibit signs of the tidal influence of Jupiter on their surface geography. By the 19th century time-predicting machines had been invented and it is now possible to accurately predict high and low tides for most coastal areas of the globe. Furthermore, tides are not limited to the world's oceans. Any large body of water, for example the Great Lakes of the United States, exhibits some tidal fluctuations in surface levels. In addition, as the Earth moves between the sun and moon the gravitational force from these objects distorts the land surface of the planet, a phenomenon called an Earth tide. However, the study of ocean tides in the 17th century does serve as an indicator of the progression of scientific thought during the century. As science in the century progressed from the introduction of the experimental approach early in the century to the investigation of the mathematical basis of natural phenomena later in the century, the individual achievements of multiple scientists began to accumulate. Finally, the science of the century had progressed to a point at which Newton could synthesize a workable solution to a problem that had puzzled scientists for over 2,000 years.

Selected Bibliography

Bonelli, Fredrico, and Lucio Russo. "The Origin of Modern Astronomical Theories of the Tides: Chrisogono, de Dominis and Their Sources." *British Journal for the History of Science* 29 (1996): 385–401.

Newton, Isaac. *The Principia.* Translated by J. Bernard Cohen and Anne Whitman. Berkeley: University of California Press, 1999.

Taton, Rene, ed. *The Beginnings of Modern Science.* New York: Basic Books, 1958.

Optics (1604–1704): The science of optics is the study of the physical properties of light. During the 17th century an interest developed among physicists to define the physical properties of light. The examination of the physical characteristics of light was not unique. Throughout the century scientists were simultaneously investigating the nature of sound, magnetism, and gravitation, to name a few. In each of these cases scientists became convinced that the prevailing views of nature did not adequately explain their observations of the real world and set forth to explain these phenomena using logic and mathematics. Such was the case with the study of optics. The study of optics during the 17th century involved not only the study of the refractive and reflective properties of light, but also the speed of light (see LIGHT SPEED), the origin of color (see LIGHT SPECTRUM), and the physical characteristics of light (see LIGHT WAVES).

The study of optics prior to the century was founded on the science of the ancient Greek mathematician Euclid. Euclid did not invent the study of optics; in fact, generations of Greek scientists before him had investigated the science of light. Rather, Euclid was a historian who collected and summarized most of the Greek work in science up until his time. As a mathematician he developed proofs for many of these theories, and for this reason his 4th-century-B.C.E. book *Optics* remained the authority on the science for centuries. In this book Euclid details Greek views on vision as well as their early studies of reflection and refraction. Some important work on the angles of reflection and refraction were also performed by the 2nd-century scholar Ptolemy. The years between the ancient Greeks and the Renaissance were not with-

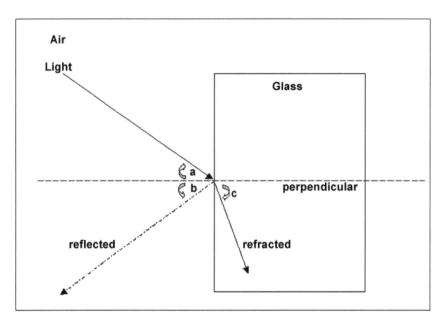

Figure 19. The principles of reflection and refraction of a light wave. In this diagram the angle of incidence (*a*) of the incoming light wave is equal to the angle of reflection (*b*). Angle *c* represents the angle of refraction. Snell's Law established the relationship between angles *a* and *c* of this diagram.

out scientific achievement. The Arabic scholar Ibn al-Haytham (also known as Alhazen) did some important work on optics in the 11th century. The 13th-century English scholar Roger Bacon also made some minor contributions to the science. However, owing to its lack of experimental proof, little of this early work would have an influence on the study of optics during the 17th century.

It was during the 17th century that some of the most important work on the interaction of light with a medium was performed. In optical terms reflection is the redirection of a light wave when it strikes the boundary between two mediums. It is different than refraction in which a portion of the light wave passes through the medium but in the process the angle of the wave is changed (see Figure 19). One of the first 17th-century scientists to apply optics to his work was the Italian astronomer Galileo. Galileo is credited with the invention of the first reflective telescope (see TELESCOPE), although he probably got his ideas on the bending of light waves from the prior invention of the microscope by Hans Lippershey and Zacharias Janssen (see MICROSCOPE). However, Galileo's telescopes were primitive and plagued by chromatic (color) and spherical (focusing) aberrations. These problems were caused by imperfections in the early lenses that resulted in the light being reflected incorrectly to the optical piece of the instrument (see TELESCOPE). The search for the cure to these problems played an important part in 17th-century optical studies.

One of the greatest contributors to optics in the early 17th century was the German astronomer Johannes Kepler. Much of his work in this area is described in his 1604 book *Ad Vitelionem Paralipomena.* Kepler contributed to the study of light waves by determining that they obeyed an inverse-square law, which means that they decrease in intensity with distance (see LIGHT WAVES), but his most important contributions came in the study of refraction. Kepler experimented with the relationship between the angle of incidence, or the angle of the incoming light from a plane perpendicular to the reflective medium (see Figure 19), and the angle of refraction of the light as it passed through the medium. His experimental approach was one of the first in the field, and involved passing a light from a common source through a transparent object and measuring the angle of the unrefracted and refracted beam. While Kepler was unsuccessful in developing a firm mathematical relationship between the angle of incidence and the angle of refraction, it was an important first step for mathematical optics. However, his understanding of optics did find application in astronomy. In order to correct the spherical aberrations of early telescopes, Kepler invented a longer version to better focus the image (see TELESCOPE). He also made a number of contributions to the study of vision, specifically how the mind detects an image (see HUMAN PHYSIOLOGY).

The next step in the study of optics was to expand on Kepler's work and determine whether a mathematical relationship could be established between the angles of incidence and refraction. The person who is credited with first developing this was the Dutch mathematician Willebrord Snell (1580–1626). Centuries earlier, Euclid had stated the angle of refraction is equal to the angle of incidence. The Greek philosopher and astronomer Ptolemy further supported this idea with his own work several centuries later. However, these relationships were not exact and did not hold true for all angles. Around 1621 Snell determined the trigonometric relationship between the angles of incidence and refraction and the distance that the light was bent from its path as it passed through a medium, called the refractive index (see Figure 19). This relationship is now commonly called Snell's Law, even though Snell did not publish his work. This law basically states that the distance light travels after being refracted through a medium is proportional to the distance it would have traveled if it had not been refracted. The French mathematician René Descartes worked out a similar relationship in his work on the effects of medium density on the motion of light. Unfortunately, Descartes thought that the nature of light caused it to speed up in denser mediums (see LIGHT WAVES). But his mathematical conclusions did indicate that a relationship exists between the angle of refraction and the speed of the object. The French mathematician Pierre Fermat would be the first to mathematically demonstrate that by using Snell's Law the speed of an object should be faster in a less-dense medium, such as a vacuum. The experimental proof of this was not possible until the 19th century when the speed of light in water and air could be measured with a higher degree of accuracy (see LIGHT SPEED).

Later in the century the mathematician Francesco Grimaldi made an interesting discovery regarding light. Grimaldi performed a number of experiments with light, some of which contributed to a better understanding of both the wave nature of light and the nature of color (see LIGHT SPECTRUM; LIGHT WAVES). In the area of optics Grimaldi was the first to describe another property of light, that of diffraction. In modern optics diffraction is the bending of light around objects. Grimaldi discovered that when a light was passed through two openings, the second of which is larger than the first, the beams of light covered a greater area than could be explained by light traveling in straight lines from the source. Thus, the light rays were being diffracted around the openings. In a second experiment (see LIGHT WAVES for figure), he placed an object in front of the light beam and noted that the shadowed area that was produced by the object did not correspond to the size predicted if the light waves traveled in straight lines. This example of diffraction led Grimaldi to believe that light behaved as a wave, and the ripples in the waves allowed it to bend around the object (see LIGHT WAVES).

One of the more interesting developments in the study of optics during the 17th century involved a description of double refraction. In double refraction a beam of light entering a medium is split into two rays. One of these rays, the ordinary ray, is refracted according to Snell's Law while the second ray, called the extraordinary ray, continues along the initial path of the light beam. The Dutch mathematician and scientist Christiaan Huygens is credited with making some of the earliest attempts to describe this phenomenon. Huygens did not discover double refraction. Instead it was first described by Danish scientist Erasmus Bartholinus (1625–1698) using crystallized calcite (calcium carbonate) as a medium. Previously, Huygens had developed a theory on how light moved through transparent objects that described light as a wave as it emits from its source. Huygens' interest in double refraction stemmed from the fact that it directly contradicted his philosophy. Huygens considered that double refraction must be a property of the crystal and not an inherent characteristic of light. He studied a large number of crystals from a geometric perspective, but was unable to develop a theory that explained the process. Isaac Newton used this as evidence that the wave theory was invalid, thus supporting his concept of a particle nature for light. Unfortunately for Newton, the process of double refraction could also not be explained by particle, also called corpuscle, theory. Since scientists of this time regarded light to be either a particle or a wave, while it actually behaves as both, it was not possible at this time to explain this phenomenon completely.

The last great contributor to 17th-century optics was Isaac Newton. Newton's 1704 publication *Opticks* summarized most of the advances in this discipline during the 1600s. Newton's own interest in optics focused mostly on the characterization of light as a particle (see LIGHT WAVES) and determining the nature of color (see LIGHT SPECTRUM). However, both of these areas involved investigations into the processes of refraction and reflection. Newton used the refractive properties of a prism to demonstrate that color is a property of light (see LIGHT SPECTRUM for figure) and not of the object being illuminated by the light. Probably Newton's most important contribution in this field was to use his studies of color and reflection to design a telescope that reduced the problem of chromatic aberration inherent to earlier 17th-century telescopes (see TELESCOPE). The principles behind a Newtonian reflective telescope are still used in modern astronomy and are incorporated into some of the world's largest telescopes.

The study of optics was an active area of research in the 17th century. The majority of the leading physicists during this time examined the properties of light from some perspective. As one of the more visible forces of nature, and one that had a long history of interpretation, light was a useful model for scientists to test their developing theories on nature. The study of optics also allows historians to visualize the progression of scientific thought in the 17th century. As the century progressed, the simultaneous development of scientific instrumentation, for example the telescope, and mathematics allowed physicists to explore light in greater detail. While many of the theories and laws designed during this time would be refined by later generations of scientists, the principles that the 17th-century investigators established were for the most part fundamentally sound. Later inventions such as cameras and binoculars are due in part to the 17th-century advances in the field of optics. The investigations of light in the 17th century represented a major advance for the science of physics in general, and the methodology used in these studies found application in the study of other physical forces of nature in the following centuries.

Selected Bibliography

Park, David. *The Fire within the Eye: A Historical Essay on the Nature and Meaning of Light.* Princeton, NJ: Princeton University Press, 1997.

Spangenburg, R., and D. K. Moser. *The History of Science from the Ancient Greeks to the Scientific Revolution.* New York: Facts on File, 1993.

Wolf, A. *A History of Science, Technology and Philosophy in the 16th and 17th Centuries.* New York: Macmillan, 1968.

P

Pendulum Clock (ca. 1600–1670): The relationship between astronomical observations and timekeeping is an ancient one. Early civilizations found it important to be able to predict the passing of the seasons by the phases of the moon and the position of the sun in the sky. Today, massive stone structures called obelisks serve as reminders of early attempts to predict seasonal changes. These structures were often strategically placed in order to predict the longest and shortest days of the year. The most famous example is Stonehenge, located in southern Britain. The masonry that survives today was built around 2550 B.C.E., although there is evidence that the current structure was built on the foundations of older works. On a daily time scale, obelisks could be utilized as massive sundials to divide the day into smaller units of measurement. Obelisks, however, were practically useless for night-time astronomical observations. As cultures and religions became more advanced, there developed a need for more elaborate timekeeping mechanisms.

For many cultures, especially the Chinese and Arabic, water clocks were developed as a more accurate mechanism of keeping time. These instruments, called clepsydras, may have been in use as early as 1500 B.C.E. By the era of the ancient Greeks (ca. 325 B.C.E.), water clocks were a popular form of timekeeping. The construction of a water clock was relatively simple. A metal or stone bowl with a small hole in the bottom was placed in a pool of water. Incremental markings on the inside of the container recorded the passage of time as the container filled. The Chinese constructed elaborate clepsydras, one of which was over thirty feet in height, which included globes and other mechanisms for measuring astronomical time. Other water clocks functioned using dripping water for the measurement of time. While superior to sundial technology, the primary disadvantage of these instruments is that they lacked a method of standardization that made precise measurements and comparison of measured time difficult.

Around the 1300s the first mechanical clocks were being constructed in Europe. These large devices utilized a series of weights and levers to measure the passage of time. The downward pull of gravity on the weight would be transferred by means of a mechanical mechanism, called an escapement, to a counting device. There is evidence that the ancient Chinese cultures also developed such devices prior to their invention in Europe. The complicated nature of these devices meant that factors such as friction between the parts, ambient temperature, and age of the device would all have an impact on the accuracy of the timepiece. It has been estimated that some erred by up to thirty minutes per day. The changes in scientific methodology, with its emphasis on experimentation (for example, see LAW OF FALLING BODIES), required a more accurate method of recording the passage of time.

The Italian scientist Galileo had many suggestions on how to more accurately measure time. He experimented with measuring time in a variety of ways, including using his own heartbeat to determine elapsed periods of time. He also suggested using the orbits of the four moons of Jupiter as a form of astronomical clock (see JUPITER). Legend has it that Galileo was inspired to invent a pendulum clock while watching the motion of the chandeliers in an Italian cathedral. While watching the pendulum clock Galileo noted that the amount of time required for the chandelier to complete one swing was constant. While the distance that the chandelier covered in a single pass would decrease over time, the period of the swing would remain the same. Intrigued by this idea, Galileo devised a series of experiments in which he varied the weight at the end of the pendulum and the length of the string. He noted that the length of the pendulum, not the weight, influenced the period of the swing. In other words, long strings took longer to complete the arc of the pendulum. By varying the length of the string, it would be possible to construct instruments that could measure precise units of time. One of Galileo's students, the French mathematician Marin Mersenne, calculated that a string length of 39.1 inches would take one second to complete the arc. Although Galileo did not actually invent the pendulum clock, he predicted that the instrument would be invaluable in measuring small units of time accurately.

Astronomical studies of the early 17th century were almost totally observational and focused primarily on describing the newly discovered complexity of the heavens. However, as the century progressed, the telescope evolved into a more complex instrument (see TELESCOPE). Astronomers were now attempting to predict orbits, periods of revolution, and distances. With this increase in complexity arose a need for more accurate time measurements. The use of a pendulum to measure time was the next step in clock evolution. However, the pendulum design as described by Galileo was not perfect. When allowed to swing freely, a pendulum does not adhere to an arc along a straight line. Instead, the pendulum swings in a circular motion owing to the motion of the Earth beneath the pendulum. Because of this the period of the swings is not always equal.

Christiaan Huygens realized that in order to keep the period of the swing constant, the pendulum would need to swing through a very small arc. By doing so, it would reduce the error associated with long arcs. In 1656 he constructed an instrument in which the arc of the pendulum was confined to a small area, called a cycloid. As the pendulum swung, the motion of the pendulum would control the rate of fall of a weight. The falling weights would in turn restore some energy back to the pendulum to compensate for losses due to friction. There were several modifications to the original design, but Huygens soon had an instrument that was accurate to less than a minute a day. Later improvements in clock design, namely the lengthening of the pendulum, would enable measurement of time down to seconds.

The development of the pendulum clock had an immediate impact on astronomy in the later decades of the century. More precise time measurements made it possible to make more accurate measurements of the planets. The pendulum clock also revealed the fact that our apparently orderly solar system had significant irregularities brought on by the elliptical nature of the planet's orbits. Using more accurate time measurements, the French astronomer Giovanni Cassini was able to record the passage of surface features on both Mars and Jupiter and then calculate the period of rotation for both of these planets. The increased accuracy of time measurements, coupled with the invention of astronomical measuring instruments (see MICROMETER), enabled some of the first investigations into discovering the speed of light (see LIGHT SPEED).

The benefits of the pendulum clock were not confined to astronomical time measurements. Since the pendulum clock is powered by gravitational forces, it could be used as a measure of gravitational intensity. If a pendulum were designed with a period of one minute at sea level, then as one moved further from the Earth, the force of gravity should weaken and thus the period should increase. Isaac Newton explained this as the inverse-square law of gravitational attraction (see GRAVITATION). Expeditions near the equator had noted that the period of the pendulum was slower at the equator than at northern latitudes, even if both readings were made at sea level. Newton suggested that this was due to the fact that the diameter of the Earth decreased as one moved north or south from the equator, forming an equatorial bulge. However, it was not until the next century that accurate measurements would be made and Newton proved correct.

While the pendulum clock was initially invented for the benefit of astronomers, it quickly found application to the general public. The grandfather clock, developed around 1670, provided accurate time measurements to the general public. For the first time in history, individuals could measure time with an accuracy of seconds, not fractions of an hour. Furthermore, since the pendulum was built on physical laws that were standard across the globe, measurements of time could be compared between locations with a minimum of error.

Unfortunately the pendulum clock was not useful for the maritime industry since its operation was impeded by the motion of the sea. This would be remedied by the invention of the **chronometer** in 1728 by the English inventor John Harrison (1693–1776). The production of the pendulum clock in the 17th century stimulated the invention of personal timekeeping devices such as the mainspring watch and the later quartz watches. For astronomers, the pendulum clock signaled the beginning of the quantitative analysis of the solar system. While observational astronomy remains active to this day, the accurate measuring of time resulted in a new generation of physicists and astronomers interested in ascertaining the physical laws of the universe.

Selected Bibliography

Hall, A. Rupert. *From Galileo to Newton.* New York: Dover Publications, 1981.

Ronan, Colin. *Science: Its History and Development among the World's Cultures.* New York: Facts on File, 1982.

Wolf, A. *A History of Science, Technology and Philosophy in the 16th and 17th Centuries.* New York: Macmillan, 1968.

Phosphorus (ca. 1670–1680): Chemistry during the 17th century was in a state of transition. Until the time of the Renaissance scientists had followed the Aristotelian philosophy that all mater was composed of only four elements—fire, water, earth, and air. Since it was the ratio of these four elements that determined the type of matter, it was theoretically possible to convert matter between forms simply by altering the ratios of the primary elements (see CHEMISTRY). Until the time of the 17th century, most studies of matter focused on the transmutation of matter. The scientists who engaged in these studies were called alchemists. However, in the 17th century there arose a series of challengers to the four-element theory of Aristotle. Leading this revolution was the English scientist, and first true chemist, Robert Boyle. Boyle's study of air (see BOYLE'S LAW; GASES) and its composition led to the birth of the scientific discipline called chemistry, a science that focuses on the properties of matter (see CHEMISTRY). As a product of this revolution, early chemists actively searched for physical evidence of elements in nature. In this process the first new element to be identified was phosphorus.

Around 1670 the German alchemist Hennig Brand (?–ca. 1670) was researching compounds to be used in the transmutation of metals to gold. He believed that he could obtain gold from human urine. After collecting human urine in a container, he allowed the liquid to partly evaporate and then distilled the remaining material. When finished, he had isolated a white, waxy material from the urine. This compound was unique in that it possessed a luminescence that did not require an initial exposure to a light source. Brand had discovered *phosphorus*, although he did not initially reveal his discovery to outside sources. At around the same time another German alchemist, Johann Kunckel (ca. 1630–1703), isolated a similar material by using a chemical reaction in which he calcified the residue from evaporated urine. While it is highly likely that Kunckel had prior knowledge of Brand's discovery, and thus was not actually a coinventor, the highly secretive nature of early 17th-century alchemy makes it difficult to establish the exact series of events.

The first scientist to identify phosphorus as an element with unique chemical characteristics was the English chemist Robert Boyle. As noted, Boyle played an important role in the demise of Greek theories of the elements (see BOYLE'S LAW; CHEMISTRY). It appears that Boyle first learned of phosphorus from a presentation of Kunckel's work in England. However, by 1680 Boyle had developed his own method of obtaining phosphorus from human urine. While other natural sources do exist, they would not be identified or accessible until the 18th century. In Boyle's method, the water in the urine was removed by evaporation, which resulted in a viscous mixture of organic material and minerals. When this was heated in the presence of sand, it was possible to separate and collect the elemental phosphorus. Boyle not only had developed a better method of obtaining phosphorus, but he also conducted a number of studies on the chemical properties of the element. In fact, it was Boyle who first correctly identified many of the properties associated with the glow noted by Brand and Kunckel. Boyle determined that phosphorus required the presence of air to glow and that the element reacted freely with the air to produce a strong chemical odor. He noted that when phosphorus was exposed to water, the result was a liquid that, when heated, produced flashes of light. This liquid suspension of phosphorus is called phosphoric acid.

While it may be considered that phosphorus was discovered in the 17th century, its discovery did not have a large impact on the science of the times. First of all, at the time of the 17th century the science of chemistry was in its infancy with the majority of the scientific effort being applied to iatrochemistry and disproving the prevailing alchemist views (see CHEMISTRY). Furthermore, even with Boyle's improved method of isolating phosphorus, it remained difficult and expensive to obtain large quantities of pure elemental phosphorus that were needed for chemical analysis. However, by the 18th century techniques had become improved. Around 1771 the French chemist Antoine Lavoisier used phosphorus in the development of his theories on the process of combustion (see COMBUSTION) and by the end of the century phosphorus was being used in a wide variety of chemical processes.

Modern chemists and geologists recognize phosphorus as an abundant mineral in the Earth's crust with over 190 known minerals containing some elemental phosphorus. In modern science, phosphorus is recognized on the periodic table of the elements by the symbol *P*. Phosphorus exists naturally as a solid, with an **atomic weight** of 30.9738. Chemists now recognize three forms of phosphorus—red, white, and black. Each of these differs slightly in chemical reactivity, with white phosphorus being the most chemically reactive at room temperature. In the biological sciences phosphorus is an important element in the process of cellular signaling and in the formation of a large number of biomolecules, including the genetic material DNA. Radioactive **isotopes** of phosphorus are used in DNA sequencing and the study of biochemical pathways. The most common use of phosphorus in society is as a component of matches. While it may be a minor note in the history of 17th-century science, the discovery of phosphorus in that century does mark an important milestone in the early beginnings of the study of matter and the final discrediting of Greek views on alchemy.

Selected Bibliography

Krebs, Robert E. *The History and Use of Our Earth's Elements: A Reference Guide.* Westport, CT: Greenwood Press, 1998.

Wolf, A. *A History of Science, Technology and Philosophy in the 16ᵗʰ and 17ᵗʰ Centuries.* New York: Macmillan, 1968.

Plant Morphology (ca. 1671–1682): Botany is the branch of science that investigates the nature of the members of the plant kingdom. Morphology is the study of the structure and function of living things. The study of plants can be said to have originated as a form of early medicine and was further advanced once early humans abandoned nomadic living and began organized agricultural practices. However, the scientific study of plant structure and physiology has been a relatively recent development for botanists. Dating from the time of the ancient Greeks, as is evident in the writings of Aristotle and Theophrastus, early botanists concerned themselves primarily with the classification of plants (see BOTANICAL CLASSIFICATION). Studies of this sort were not limited to Western cultures as Chinese botany was paralleling the work being done in Greece. But work into how plants functioned was limited in both cultures. Theophrastus, considered by many to be the father of Greek scientific botany, did some important work on plant morphology. While Theophrastus did advance the study of botany by developing standardized terms and descriptions of plants, it appears that little knowledge of his work endured into later periods. Thus, he probably had minimal influence on subsequent generations of botanists. In the interim period between the time of Theophrastus and the 17ᵗʰ century, most of the advances in botany involved the identification and use of plants with medicinal properties. While the 18ᵗʰ century is often regarded as the beginning of the true research into plant morphology and physiology, there were several important discoveries made during the 17ᵗʰ century that set the stage for these later works.

In the late 1600s there were two primary investigators of plant morphology, the Italian biologist Marcello Malpighi and the English botanist Nehemiah Grew (1641–1712). The invention of the microscope in the 17th century had revolutionized the study of biology (see MI-CROSCOPE) and both of these men belonged to a new generation of scientists who made extensive use of this invention. They simultaneously conducted research on plant morphology, and although the two did not appear to directly collaborate in their efforts, in several important areas their work frequently overlapped. Their individual publications on plant morphology, Malpighi's *Anatomy Plantarum* (1675) and Grew's *Anatomy of Plants* (1682), are both considered as landmark publications in the field of plant morphology. Malpighi was a skilled microscopist for his time. As one of the original microscopists, he made significant contributions to a number of areas of biology, including anatomy, the circulation of the blood, and preliminary studies on insect anatomy (see BLOOD CIRCULATION; ENTOMOLOGY). Grew successfully applied his training as a physician and knowledge of animal anatomy toward scientific studies of plants. The close parallels of their findings and subsequent publications makes it difficult to determine the originator of many of their discoveries. However, together they made a significant impact in their field.

Under the microscope both Malpighi and Grew noticed that plants consisted of numerous individual compartments, or *utriculi* as Malpighi called them. These were in fact cells, the first of which had been described by Robert Hooke a few years earlier (see CELLS). While Hooke had observed dead cells, Malpighi and Grew worked with living tissue, which enabled them to make important discoveries on the microscopic morphology of plants. They did not recognize the cells as living units; instead they both thought of them as locations for sap storage. However, they did note their apparent importance to the plant and the fact that they varied in structure in the different tissues of the plant. For example, Grew noted that the **xylem** of the plant, the structure that transports water, is greatly elongated in comparison to other plant cells. The modern botanical term **parenchyma**, which signifies undifferentiated plant cells, originated with Grew's descriptions of cell types. Malpighi diagrammed many of these cell structure differences in one of the later volumes of *Anatomy Plantarum*.

Probably one of the most important advances that these men made to botany was the establishment of a new method of identifying morphological function. Traditionally the classification of plant structures was made based primarily on the function that the structure provided for the plant. For example, roots grew downward and transported food from the earth. However, by their efforts Grew and Malpighi established the practice of defining morphological structures by multiple criteria, including how the structure originated in the young developing plant, the anatomy of the part (often including microscopic investigations), and the role of the structure as determined by the process of experimentation. An excellent example of this is Grew's early experiment with **geotropism**. While many believed that roots originated by contact with the earth, Grew designed an experiment in which seeds were germinated in humid air in the absence of soil. The roots of these plants still formed and grew downward while the stems grew upright, thus indicating that the soil was not the contributing factor in root formation. Furthermore, from an embryonic perspective they noticed that the roots originated at the opposite pole of the germinating seed from where the stem developed. Thus, the plant was being influenced by a physical force (in this case gravity) rather than the presence of specific nutrients such as air and water. Malpighi made an initial prediction on the relationship of leaves to plant growth by cutting leaves from young plants and observing the impact on the production of fruits and seeds. Although the role of leaves in **photosynthesis** would take some time to develop, the decrease in fruit size indicated to Malpighi that these structures had a role in plant nutrition.

Together these scientists made a number of other important contributions. Grew noted that when a tree grows thicker by a process now called secondary growth, it is in response to changes in the **cambium** layer located just under the bark of the tree. Grew also noted that it was possible to chemically isolate the green pigments from a plant, and described some early chemical properties of these photosynthetic pigments called chlorophyll. They discovered that the growth of a shoot of a plant occurred at the buds and then proceeded to make some early microscopic descriptions of bud structure. Together they also made a number of discoveries concerning the structure of the flower and the process of plant reproduction. Grew proposed that the **stamen** of the plant functioned as the male reproductive organ, while the **pistil** was the female section of the plant. Finally, Malpighi independently made an important discovery when he related the trachea, or respiratory structures, of insects (see ENTOMOLOGY) to the structure of spiral vessels between the cells of plants (see Figure 20). He suggested that like animals, plants conduct **respiration**. Although ahead of his time, Malpighi had established the possibility that all living things shared certain characteristics, and that respiration may be one of them.

The fact that two independent researchers conducted such closely related work at practically the same time during the 17th century is a

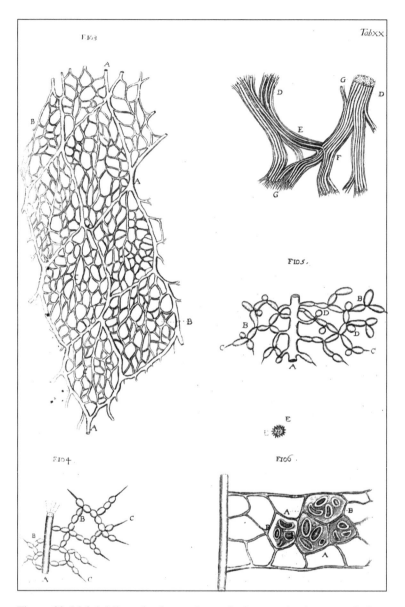

Figure 20. Malpighi's early observations of microscopic plant morphology from a number of plant tissues. (From *Anatomy Plantarum* [1675], Library of Congress Photo Collection.)

strong indicator that the time was ripe for such discoveries to occur. Both men were excellent botanists with fine observational skills. Their discoveries were enhanced by the fact that they also possessed training in a number of scientific disciplines. Aided by 17th-century advances in scientific instrumentation and a new method of scientific thinking based on logic, Grew and Malpighi were able to establish several important foundations for later studies on plant morphology that persisted into the 18th century. Their discoveries did not have a tremendous impact on the science of the 17th century. However, in the 18th century the English scientist Stephen Hales would continue the work of Grew and Malpighi by applying his knowledge of physical science to the study of plant morphology and physiology. This led to development of experimental botany, a field that persists in modern science in the improvement of agriculture and the development of pharmaceutical medicines from plants. As with many of the scientific discoveries that occurred following the Renaissance, the advances that occurred in the 17th century served as a transition to a new form of science based on experimentation and logic.

Selected Bibliography

Morton, A. G. *History of Botanical Science.* London: Academic Press, 1981.
Serafini, Anthony. *The Epic History of Biology.* New York: Plenum Press, 1993.

Probability (1654–1657): There were several important developments during the mathematical revolution of the 17th century. The invention of logarithms, analytical geometry, binomial theory, and calculus all followed a logical series of events as mathematicians attempted to develop new mathematical procedures to facilitate the advances in science during their times. For example, logarithms allowed the manipulation of large numbers for the study of astronomy and physical science (see LOGARITHMS), while calculus gave the ability to examine rates of change in relation to time (see CALCULUS). Each of these discoveries typically built on the work of the great ancient Greek mathematicians. However, the discovery of the laws of probability was unique. The study of probability involves the ability to predict the chances or frequency that an event, or series of events, will occur. While it is likely that the great ancient mathematicians were aware of chance events, it does not appear that they made any attempts to develop mathematical principles to explain them. Thus the study of prob-

ability can be said to have had its origins around the time of the 17th
century. Furthermore, while the other advances were all initially developed to further the investigations of the scientific revolution, the
foundation of the laws of probability actually had its origins as a
method of explaining games of chance, and not as an aid to a specific
scientific discovery. Still, once established, these rules would become
an important tool for scientists in their studies of the physical world.

The mathematical study of probability is believed to have originated
with what was called the *problem of the points*. First described by the
Italian Luca Pacioli (ca. 1445–ca. 1509), although it was known of earlier, this problem involved how to proportionally divide up the stakes
between the contestants of an interrupted game if one knew the score
of the contestants and the final score needed to win. The Italian mathematician Gerolamo Cardano (1501–1576) went a step further and
published a number of publications on probability and games of
chance, specifically those involving dice, during the 16th century. Cardano was one of the first to use the multiplication rule for probabilities
that involve independent events. For example, if when flipping a coin
the chances of getting a head are 1 in 2 ($\frac{1}{2}$), then the chances of
getting heads on two independent flips of the coin would be 1 in 2
($\frac{1}{2}$) times 1 in 2 ($\frac{1}{2}$) or 1 in 4 ($\frac{1}{4}$). At about the same time similar
studies were being made by the mathematician Niccolo Tartaglia (ca.
1499–1557). However, despite their work there did not exist a formal
mechanism of addressing probability at the start of the 17th century.

Since probability is widely in use in modern science, and there was
a wealth of new information being discovered in the early decades of
the 17th century, it would be easy to think that laws of probability were
being developed as a result of these developments. However, this was
not the case. While some of the primary mathematical minds of the
17th century contributed to the study of probability, including Blaise
Pascal, Pierre de Fermat, and Christiaan Huygens, once again they
were drawn to the problem as a means of explaining games of chance,
and not initially for any specific scientific reason. It is believed that
Pascal and Fermat were directed to the study of probability theory by
the French knight and renowned gambler Chevalier De Méré, who
wanted to know the chances of certain rolls of the dice. De Méré, a
friend of the King of France, recognized the importance of probability
in games of chance. In response to specific questions posed by de
Méré, Pascal consulted with Fermat. While their consultations did not
produce fruitful results for de Méré, the collaboration between the two
mathematicians did serve as an important catalyst for additional de-

velopments in probability. Specifically, both men became interested in how to accurately solve Cardano's problem of points. Pascal proposed that the winnings be divided among the contestants proportional to their probabilities of winning the game. To assist in this calculation, Pascal designed an arithmetic triangle to calculate the chance of winning. This triangle was the same used for Pascal's work with the binomial theorem to determine the coefficients of expansion (see BINOMIAL THEOREM for figure). Fermat proposed a similar method, but focused his calculations on the probability of a contestant winning the entire match. What is important is that in both cases attempts were being made to mathematically define probability, although a consensus had once again yet to be reached.

The primary attempt at a single unifying theory of probability in the 17th century was made by Christiaan Huygens. While Huygens was aware of the work of both Pascal and Fermat, it appears that he developed his theories of probabilities independently of the French mathematicians. Huygens was also one of the first to recognize the power of the probability theorem outside of games of chance. From this work he developed a process called mathematical expectation that involves generalizing the arithmetic **mean**, a useful function in business where it may be used to determine mean prices and profit levels. This was a precursor to modern statistical analysis. Furthermore, Huygens' work included mathematical equations to define the probability or chances of certain events occurring. His work remained the primary authority on probability theory until the publication of *The Art of Conjecturing* in 1713 by Jakob Bernoulli. Bernoulli's work contained references to Huygens' investigations on probability. Additional work in the 18th century was conducted by the French mathematician Abraham de Moive (1667–1754), whose book *Doctrine of Chances* (1718) discussed probability as it related to mortality and not the traditional subject of games of chance. This is the same type of probabilities used by insurance companies to calculate premiums based on the age of the applicant.

The discovery of probability did not have a tremendous impact on the science of the 17th century. Scientists of this time were not as interested in determining the chances of an occurrence as they were in the rate of a natural event or manipulation of a geometric figure. However, while the work of Pascal, Fermat, and Huygens in developing a probability theory may appear to be secondary in comparison to their other accomplishments during the 17th century, the initial investigations that they performed would have an influence on the next generation of mathematicians. Probability analysis and its close relative statistics are now a prime component of scientific research in almost any discipline. Although many of the other achievements of the 17th century can be said to have been built on the work of centuries of scientists, it can be said that probability originated during the time of the 17th-century scientific revolution.

Selected Bibliography

Cooke, Robert. *The History of Mathematics: A Brief Course.* New York: John Wiley & Sons, 1997.

Katz, Victor J. *A History of Mathematics: An Introduction.* Reading, MA: Addison-Wesley Longman, 1998.

Maistrov, L. E. *Probability Theory—A Historical Sketch.* New York: Academic Press, 1974.

Stillwell, John. *Mathematics and Its History.* New York: Springer-Verlag, 1989.

Swetz, Frank J. *From Five Fingers to Infinity: A Journey through the History of Mathematics.* Chicago: Open Court Publishing, 1994.

S

Saturn (1610–1656): Of the nine planets, four are visible in the night sky without the use of a telescope. Venus, Mars, Jupiter, and Saturn were collectively called the wanderers by the ancient Greeks owing to their sometimes-erratic motions in the night sky. Since these motions were distinct from the predictable patterns of the moon, sun, and stars, these planets were considered to be representatives of the gods. As such they became the foundations for mythology. The faintest of the planets, Saturn, represented the god of agriculture by the ancient Greek, and later the Roman, civilizations.

For most of recorded time the known universe was considered to consist of stars and six heavenly bodies—the sun and moon and the four planets. The invention of the telescope in the 17th century (see TELESCOPE) gave astronomers the ability to observe these objects in greater detail. While Jupiter and Mars would serve as a testing ground for emerging theories on the physical structure of the solar system (see JUPITER; KEPLER'S THEORY OF PLANETARY MOTION; MARS), the dimness and slow-moving nature of Saturn would confine its discoveries primarily to the field of observational astronomy. Galileo was one of the first to observe Saturn using one of the early refracting telescopes. When Galileo first turned his attention to Saturn in 1610, he noticed that the normal spherical planetary disk was marred by small projections on either side. The primitive nature of the early telescopes inhibited Galileo from making more detailed observations, but Galileo proposed that what had historically been considered the planet Saturn was actually a planetary system consisting of three planet-like objects in close proximity to one another. However, unlike the other planets, the shape of Saturn appeared to change over time. In 1612 he observed that the three objects had merged into a single planetary disk. Unable to solve this problem, Galileo instead focused on studying the nature of Jupiter's moons (see JUPITER).

Early refracting telescopes were plagued by both focusing and chromatic (color) problems (see TELESCOPE). It would take several decades and improvements in the imaging system of the early telescopes before Saturn could be observed in enough detail to determine that Galileo's planets were actually rings. By the mid-1600s Christiaan Huygens had developed an improved method of grinding optical lenses. When these were placed into the next generation of telescopes, the result was a significant improvement in image quality. Using one of these refined instruments, called a long telescope because of its increase in overall length, Huygens turned his attention to the question of Saturn. By 1656 Huygens had determined that what Galileo had observed was not a three-planet system, but rather a single planet surrounded by a thin, flat ring. He noted that this ring did not physically come in contact with the planet, but rather orbited the planet at an inclination, or angle, of about 26.7 degrees. Furthermore, once it had been determined that the projections were actually a thin ring, it became possible to explain why the objects that Galileo was observing in 1612 had disappeared. Huygens suggested that since the rings surrounding Saturn were at a similar inclination as that of the Earth (23.4 degrees), there would be times during the Earth's orbit when we would be observing the rings head-on. Since the rings were thin, it would appear to the observer that they had vanished completely. Huygens published these observations in 1665 and in support of his claims he accurately predicted a return of the rings in 1671. We now recognize that the relationship between the orbits of Saturn and Earth results in this phenomenon occurring about once every fourteen years, a discovery made possible by the excellent observational abilities of Huygens.

With both attention and improved instrumentation focused on Saturn, the next series of discoveries would be centered on identifying any satellites in orbit around the planet. In 1665, Huygens discovered Saturn's largest satellite. This satellite would later be named Titan, the name of the class of gods to which the god Saturn belonged. Modern astronomers are interested in Titan because of the fact that it has a dense atmosphere, the only known satellite in the solar system to have such. In addition, the atmosphere of Titan may be rich in organic molecules, possibly even amino acids. Amino acids are the building blocks of proteins, important **biomolecules** of living cells. However, in the 17th century the discovery of Titan held another importance. The detection of Titan brought the total number of known satellites in the solar system to six (our moon, the four moons of Jupiter, and Saturn's Titan). This number precisely matched the number of known planets (Earth, Mars, Venus, Jupiter, Saturn, and Mercury). To Huygens, this pairing of satellite and planet number was not a coincidence. Instead it indicated a form of harmony, and he incorrectly assumed that no new satellites would ever be discovered. In fact, although equipped with some of the best telescopes of his time, so strong was this belief that Huygens abandoned the search for additional satellites. His ideas were proven incorrect in 1671 when Giovanni Cassini discovered Saturn's satellite Iapetus. Cassini would follow this with the discovery of Rhea in 1672 and with the satellites Tethys and Dione in 1684. As with Jupiter, only the larger satellites could be detected with the telescopes of the 17th century (see Figure 21). Additional discoveries would have to wait until improvements in telescope design and photographic imaging were made. Saturn is now recognized to have eighteen satellites (see Table 3), with some so small that they were only detected by the Pioneer and Voyager spacecraft in the late 20th century.

The composition and origin of the unique ring system of Saturn have been a focal point of planetary science since the time of its discovery by Huygens. While it is now recognized that the other gas giants, most notably Jupiter, all may possess similar structures, none of them remotely compare to the brilliance of Saturn's rings. In 1675 Cassini observed that the rings of Saturn were not a single object; instead there was a division between them. Modern astronomers have observed that there are actually several divisions in the rings. This major division is called the Cassini Division after its discoverer. However, Cassini considered the rings themselves to be solid objects, a mistake that may easily be attributed to the telescopes of his time. If one looks through a small telescope today, the rings of Saturn appear smooth as well.

While Isaac Newton's work on gravitation would provide the unifying force for many astronomical theories developing in the 17th century (see GRAVITATION), it would actually complicate investigations of the nature of Saturn's rings. Since the rings consisted of matter, then over time the gravitational attraction of this matter should have forced the material to coalesce into larger bodies or satellites. Yet this obviously had not occurred. Pierre-Simon Laplace, best known for his work with the nebular hypothesis of solar system formation (see NEBULA), would suggest that since the rings were at different distances from the planet, they were not all subjected to the same gravitational pull. These differences in gravitational attraction inhibited the formation of sat-

Figure 21. This photo is from a collection of images of Saturn and its moons collected by the Voyager 1 spacecraft in November 1990. The moon in the forefront is Dione. To the lower left of Saturn are the moons Enceladus and Rhea, while the moons to the right of Saturn are Tethys and Mimas. In the upper left of the image is the moon Titan, the focus of the 21st- century Cassini space probe (arrival date 2004). (Image courtesy of NASA, the National Space Science Data Center, and Dr. Bradford A. Smith.)

Table 3
Satellites of Saturn in order of discovery. Location indicates the order of the moon from the planet. In addition, in 1995 four additional moons of Saturn may have been identified, but are not yet named.

Name	Year Discovered	Location	Discoverer	Nationality
Titan	1655	6th	Christiaan Huygens	Dutch
Iapetus	1671	8th	Giovanni Cassini	French
Rhea	1672	5th	Giovanni Cassini	French
Dione	1684	4th	Giovanni Cassini	French
Tethys	1684	3rd	Giovanni Cassini	French
Enceladus	1789	2nd	William Herschel	British
Mimas	1789	1st	William Herschel	British
Hyperion	1848	7th	William Bond	American
			William Lassell	English
Phoebe	1898	9th	William Pickering	American
Janus	1966	10th	Audouin Dollfus	French
Epimetheus	1978	11th	John Fontain	American
			Stephen Larson	American
Atlas	1980	15th	Voyager 1	American
Calypso	1980	14th	Voyager 1	American
Helene	1980	12th	P. Laques	French
Pandora	1980	17th	Voyager 1	American
Prome-theus	1980	16th	Voyager 1	American
Telesto	1980	13th	Voyager 1	American
Pan	1990	18th	Voyager 2	American

ellites from the material. However, for this to occur the material within the rings could not be a solid, but rather must consist of numerous small particles. Unfortunately, these were difficult to detect. However, in 1859 this idea was confirmed by the Scottish mathematician James Clerk Maxwell (1831–1879). Maxwell's thoughts on ring composition would gather widespread support in its time, and would later be confirmed after visits to Saturn by the Pioneer and Viking space probes. From a mathematical perspective, the French astronomer Edouard Roche (1820–1883) calculated that there was a distance from the planet in which the gravitational force of the planet would tear smaller objects apart. This region, now called the Roche Limit, is a distance equal to 2.44 times the equatorial radius (the distance from the center of the planet to the equator) of the planet. In the case of Saturn, the Roche Limit is approximately 91,000 miles. The entire ring system of

Saturn falls within this limit and thus it is the gravitational effect of the planet that maintains Saturn's rings.

Exploration of Saturn continues into this century. In 2004 the Cassini space probe, the largest probe launched to date by NASA, will take a detailed look at the gas giant and its moons. On board is the Huygens probe, designed to descend into the atmosphere of Titan and sample its contents. This visit will culminate 350 years of observation that began in the 17th century.

Unlike the discoveries in the 17th century involving both Jupiter and Mars, the astronomical advances associated with the study of Saturn did not have a direct impact on the science of the times. By the end of the century satellites in orbit around planets appeared to be common in the solar system. However, the discovery of Saturn's rings and satellites did mark an important advancement for 17th-century science. These discoveries would not have been possible without technological advances in astronomical instrumentation, namely improved lenses and more powerful telescopes. Huygens himself noted that his discoveries would have been possible earlier in the century had the instruments been available. During the 17th century the development of specialized instruments for scientific research was at its infancy (see MICROMETER; PENDULUM CLOCK). While much observational work still needed to be performed, from this point forward scientific advances would usually be dependent on technological advances in instrumentation.

Selected Bibliography

Asimov, Isaac. *Eyes on the Universe: A History of the Telescope.* Boston: Houghton Mifflin, 1975.

Moore, Patrick. *The Great Astronomical Revolution: 1543–1687 and the Space Age Epilogue.* Concord, MA: Albion Publishing, 1994.

Motz, Lloyd, and Jefferson H. Weaver. *The Story of Astronomy.* New York: Plenum Press, 1995.

Wilson, Robert. *Astronomy through the Ages: The Story of the Human Attempt to Understand the Universe.* Princeton, NJ: Princeton University Press, 1997.

Scientific Reasoning (1620–1687): Since the time of Aristotle, logic has been an integral component of the scientific thinking process. The philosophers of ancient Greece based their understanding of the natural world almost exclusively on introspective thought and logic. It was during this time that logical thinking, the idea that conclusions must be supported by observation, was first applied to scientific reasoning. There are two major patterns of logic analysis. The first, called deductive reasoning, uses general information to solve specific observations. Deductive logic moves from generalities to specifics. In mathematics, an equation (a specific case) may be solved by the application of a series of general rules. For example, the equation $3x + 15 = y$ may be solved by applying the general principles of first subtracting 15 from both sides, and then dividing both sides of the equation by $\frac{1}{3}$. The opposite of deductive reasoning is the process of inductive reasoning. In inductive reasoning, specific observations are used to develop general conclusions. For example, if one samples a number of grapes, and all of them taste sweet, the inductive response would be that grapes in general are sweet, although the possibility exists that this reasoning may be proved wrong in the future. The end product of inductive reasoning reflects the probability that the event will occur, not that the result is an inevitable truth. While the deductive process finds the solution for a specific problem, the inductive process leaves room for future exploration of other variables that may influence the observation. Frequently statistical analyses, such as probability (see PROBABILITY), are used to predict the possibility of future outcomes in inductive reasoning.

The modern scientific method is a synthesis of both inductive and deductive reasoning. In this method, observations of the natural world lead to hypotheses, or tentative explanations, as to the cause of the event. Using the inductive method, scientists often propose several general explanations for the same observation. The validity of these explanations may then be determined by the process of experimentation. The experimental design should be such that the experiment tests the contribution of a single variable (or at most a limited number) to the observation in question. Often in science, the results of the experiment do not support the original hypothesis. In these cases, the information derived from the experiment is used to deduce a new, enhanced hypothesis for the observation. This process of deduction allows the scientist to form a new, more specific hypothesis from the general data. At some point, the experimental data logically supports a single hypothesis. This, however, is not the end point of the scientific method. Although a scientist may have what appears to be an explanation for the observation, the results of the experiment must be confirmed by other scientists. With repeated confirmation, these proven hypotheses may become widely accepted theories over time. The primary difference between deductive and inductive thought lies in the use of theories. In deductive thought, one starts with a theory and then moves toward observations, while in inductive thought the opposite is true. While the term *theory* often represents speculative thought in modern society, in the scientific method theories are scientific findings that are supported by a wealth of experimental data.

While the scientific method may appear to be both a practical and logical method of investigating scientific phenomena, it actually took a considerable amount of time for the method to evolve into its current form. During the time of the ancient Greeks, philosophers such as Aristotle did not employ experimentation to confirm their observations. In fact, the great Greek philosophers believed that observation alone, without the support of either experiments or mathematics, was satisfactory enough to understand the reality of the world around them.

This method of scientific thinking persisted for centuries. It was not until the Renaissance period in Europe, and a resurgence of an interest in invention, science, and discovery, that scientists began to mathematically and experimentally question the world around them. One of the first documented uses of experimentation to confirm observations was by William Gilbert in 1600. Gilbert was a prominent physician of his time who became interested in applying the principles of magnetism to medicine. In a book entitled *De Magnete* (*The Magnet*) Gilbert outlined a detailed inductive investigative approach to explain the use of magnets in medicine (see MAGNETISM). The results of his studies had far-reaching implications on science. His research on magnetism provided the foundation for studies well into the 18th century. But more importantly, he had a direct influence on the thinking of both Galileo Galilei and Johannes Kepler, two individuals who would later in the 17th century further define the scientific method.

While the basic premises of the scientific method, especially the role of experimentation, were becoming more recognized in the 17th century, the controversy persisted over whether inductive or deductive processes should form the basis of this new method of thinking. This controversy is best illustrated by examining the works of two prominent philosophers of this time, Francis Bacon (1561–1626) and René Descartes. In *Novum Organum* (1620), Bacon challenged the current philosophies of the time, which still centered on Aristotelian logic. He was concerned that the deductive method of thinking would not satisfy the needs of society. Bacon instead focused on the use of observation and experimentation, supported by inductive thinking, to examine the natural world. He believed that the laws of nature should be derived from specific observations. Bacon furthered his position with the publication (posthumously) of *New Atlantis* in 1627. This fictitious work described a futuristic research institution that focused on inductive scientific processes over Aristotelian methods thought. The concepts presented in *New Atlantis* resulted in the formation of several early scientific societies in Europe, including the Royal Society of London.

At approximately the same period deductive reasoning was becoming the prominent form of logic in mathematics. The link between scientific theories and mathematics had been proposed earlier in the century by the German scientist Joachan Junguis (1587–1657). In 1637 the French philosopher René Descartes firmly established the power of deductive reasoning in mathematics. The 17th century was a time of modernization for mathematics and Descartes was a pioneer in development of analytical geometry and number theories. In *Discourse on Method,* he challenged the use of inductive thought in scientific thinking and instead reasoned that observations could best be solved by first starting from recognized facts, and then progressing in steps toward the solution.

While the philosophical debates over the method of scientific reasoning often polarized the scientists of this time, there were individuals who recognized the strength in a synthesis of these ideas. Among the first of these was the Italian astronomer and physicist Galileo Galilei. Galileo noticed that a pendulum always completes the same number of swings in a given period of time. Intrigued by this observation, and probably stimulated by studies of Gilbert's experimental approach to magnetism, Galileo designed a series of experiments in which he altered the weight of the pendulum. His findings indicated that the length of the string, and not the weight of the pendulum, influenced the rate of the pendulum swing (see PENDULUM CLOCK). He intentionally designed his experiments so that they could be easily reproduced and verified. More importantly, Galileo had used mathematical principles to solve problems of the physical world. This combination of the deductive and inductive processes was the true beginning of the modern scientific method. Despite criticism from colleagues, Galileo continued this method of scientific inquiry in his future studies on the laws of motion (see LAW OF FALLING BODIES). However, the use of mathematics in examining the physical properties of nature was not unique. Others, such as Johannes Kepler's use of mathematics to predict the elliptical nature of planetary orbits, demonstrated that the time was right for a revolution in science (see KEPLER'S THEORY OF PLANETARY MOTION).

The final synthesis of the scientific method occurred at the hands of Isaac Newton. Isaac Newton was considered by many to be the driving force behind the scientific revolution of the 17th century. As a scientist, Newton made contributions to many diverse disciplines including the study of light (optics), studies of gravity, and the design of scientific instruments (see GRAVITATION; OPTICS). In scientific methodology, Newton built on the foundations of scientific thought presented by Galileo, Gilbert, and Bacon. In 1687, as part of a landmark physics publication entitled *Principia Mathematica*, Newton outlined the basis for scientific thought that persists to this day. The major premises presented by Newton firmly established the link between mathematics, including the new mathematics of calculus, and analytical thought. In his method, Newton did not simply support his statements with philosophical thought. Newton put forth the idea that the results of experimentation and observation should be considered to be accurate until disproved by scientific processes. By using the experimental principles set forth by Galileo, Newton used the scientific method to examine the universal laws of nature (see GRAVITATION; LAWS OF MOTION). The combination of deductive mathematics and inductive experimental reasoning proved to be a successful combination for Newton and through his success found application in all scientific disciplines.

While at first glace the development of scientific reasoning may not appear to be a groundbreaking advance for 17th-century science, in fact it represents one of the most important achievements in the history of science. The 17th century marks a transition for scientific thought as scientists and mathematicians departed from ideas and procedures established centuries earlier in favor of laws and principles based on logic, experimentation, and the use of mathematical and statistical analysis. While the use of experimentation and reasoning was a novel idea at the start of the 17th century, it proved to be a powerful tool in the course of the scientific revolution occurring during this time. By the time of Newton, few scientists continued to practice Aristotelian science. As a result, the pace of scientific discovery was accelerated, as was the quality of work produced.

Selected Bibliography

Scott, Joseph F. *The Scientific Work of René Descartes (1596–1650)*. London: Taylor and Francis, 1976.

Taton, Rene, ed. *The Beginnings of Modern Science*. New York: Basic Books, 1958.

Slide Rule (ca. 1630–1654): Beginning in the Renaissance the rapid development of scientific thinking, especially in the studies of astronomy and physics, required an improved method of performing calculations involving large numbers. Mathematics had changed little in the times leading up to the 17th century. Exponential notation was in its infancy and the only reliable calculating device in widespread use was the abacus. Invented originally by the Babylonians, the modern form of an abacus with counters strung along wires had originated around the time of the Romans centuries earlier. The publication of a paper describing the invention of logarithms in 1614 by John Napier and Henry Briggs would revolutionize the ability of scientists to perform complex mathematical calculations. Logarithms enabled scientists to multiply and divide any two numbers using the simpler mathematical processes of addition and subtraction (see LOGARITHMS). By the 1630s logarithmic tables had been constructed for most numbers from 1 to 100,000. While the tables were useful, what scientists required was a device that expedited logarithmic calculations. The result was the invention of a device called a slide rule.

Like many scientific instruments of the 17th century, claims of inventing the first slide rule were made by multiple people. Historically, simultaneous inventions are typically an indication that a culture had advanced to a level where the technology to make the discovery or invention was widespread. Such was the case with the slide rule. Around 1620 the English astronomer Edmund Gunter constructed the first instrument designed specifically for logarithmic calculations. Gunter invented a scale on which were plotted numerals in a fashion similar to a modern ruler. However, unlike a ruler where the numbers are spaced at a predetermined standard distance from one another, the distances between the numbers on Gunter's scale were proportional to the logarithm of the number. Calculations were then performed using a compass to measure the distances. Multiplication of large numbers was accomplished by first measuring the distance corresponding to the first number and then adding it to the distance of the second number. The sum of these two distances indicated the product of the two numbers. The accuracy of the device was dependent upon its size, since smaller scales had fewer divisions imprinted on them. However, some of these instruments were over two feet in length. Division was accomplished in the same manner using the differences between the two distances. While not truly a slide rule, since it possessed no moving parts, this instrument found widespread use in navigation where they were commonly called a Gunter's Scale or a gunter.

The first instruments that used moving scales to make calculations in place of measuring distances using compasses were the circular slide rules. The exact origin of these devices remains under debate. Around 1630 the English mathematician William Oughtred converted the linear Gunter's Scale into a circular device. It consisted of two concentric circles, each inscribed with a logarithmic scale. The two circles could then be manipulated by the operator to give the sum or difference, corresponding to multiplication or division operations, between the two numbers. The principles behind its operation did not differ appreciably from the Gunter's Scale, but the instruments were smaller and more easily operated. While Oughtred published his results in 1632, one of his students, the English mathematician Richard Delamain (ca. 1600–ca.1644), is claimed to have published details on the circular slide rule in 1630. Whether Delamain was actually an independent inventor of the instrument, or simply had access to Oughtred's work, is uncertain. Regardless, this instrument did not have much effect on the science of the times. While the Gunter's Scale remained popular in maritime navigation, there was little progress in the development of the circular slide rule. By the end of the century there appears to have been little interest in the device.

The first examples of a modern linear slide rule, where a portion of the instrument slides between two fixed pieces, also had its origins in the 17th century. Oughtred is credited with having invented the first rectilinear (lined) slide rule around 1621. This device was basically an improvement on the Gunter's Scale and consisted of two logarithmic rulers that were slid past one another to perform calculations (see Figure 22). Since not all numbers were indicated by markings on the slide rule, the precision of the instrument was dependent on the size of the device and the ability of the operator. The first slide rule that consisted of a single unit with sliding parts is believed to been constructed around 1654 by the inventor Robert Bissaker. However, there is some evidence that Edmund Wingate (1593–1656) may have suggested a similar instrument earlier in the century. Others made additional improvements toward the end of the century, mostly for specific applications in industry. Unlike the circular instruments, these linear devices were recognized as useful calculation aids and their use quickly spread throughout Europe by the 18th century.

The modern form of the slide rule was invented around 1859 by the French military expert Amedee Mannheim (1831–1906). The Mannheim ruler was one of the first slide rules to use an indicator for more precise positioning of the scales. This device remained relatively un-

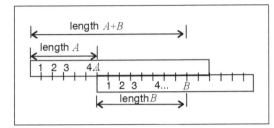

Figure 22. The use of a linear slide rule. A slide rule enables an operator to multiply two numbers by adding the logarithms of the number. Since the numbers on the slide rule are logarithmic, one can slide the rulers so that the distance *A* represents the logarithm of the first number, and the distance *B* the logarithm of the second. The total distance (*A* + *B*) represents the logarithm of the product of *A* and *B*. Using a logarithmic table it is possible to convert this number back to the common form.

changed for over a century and today Mannheim-style slide rules are still available. A glass indicator similar to ones on modern instruments was added at the end of the 19th century. Engineers and scientists of the 20th century would use the slide rule as their primary mode of calculations until the invention of the calculator in the 1960s. The mathematical capabilities of modern calculators and computers have rendered the slide rule practically obsolete, although it is still utilized as an inexpensive calculation device and teaching tool, especially in engineering. Despite the fact that the popularity of the device has waned in this century, the longevity of the instrument, spanning almost three centuries, is a tribute to its importance to the scientific community.

Selected Bibliography

Cajorie, Florian. *A History of the Slide Rule and Allied Instruments.* Mendham, NJ: Astragal Press, 1994.

Wolf, A. *A History of Science, Technology and Philosophy in the 16th and 17th Centuries.* New York: Macmillan, 1968.

Sound (ca. 1600–ca. 1660): What humans perceive as sounds are actually vibrations in the surrounding air that are detected by the human ear. While the human ear has the ability to detect approximately 400,000 distinct sounds, there are many vibrations that fall outside of the range of human hearing. Scientific interest in the physical nature of sound most likely predates the ancient Greeks. However, it is known that the Greeks expressed an interest in sound from a philosophical perspective. The Greeks were mostly interested in the study of acoustics, a branch of science that focuses on how sound is heard. The term **acoustics** is derived from the Greek word *akoustos*, which means hearing. The term has its origins in the 17th century, having been introduced by the French physicist Joseph Sauveur (1653–1716). The nature of sound was investigated in Greece as early as the 4th century B.C.E. The Greek philosopher Aristotle correctly predicted, without experimental verification, that the movement of sound from its originating source is due to the motion of the air. Unfortunately, Aristotle's philosophy was flawed when he predicted that low frequencies move at a slower rate than high frequencies. However, the Greeks were highly successful at incorporating their knowledge of sound into their architecture as Greek amphitheaters are widely recognized as having superb sound quality. This is a fine example of the application of science to art during this time. Little work was conducted on the properties of sound in the centuries following the ancient Greeks. However, during the 17th century there was a renewed interest in the physical properties of the universe. This resulted in a number of advances being made in the study of sound. These advances focused primarily on acoustics, determining the speed of sound and the relationship between the propagation of sound and air.

In the early 17th century there was a significant amount of interest in the study of acoustics. This interest centered primarily on developing a scientific understanding of music. One of the earliest investigators in the century was the renowned Italian scientist Galileo. Coming from a family with a musical background, Galileo expressed an early interest in understanding the scientific basis of sound generated from string instruments. In his 1638 publication *Discoursi*, Galileo summarized his studies on the frequency and **pitch** of sounds. The frequency of a sound is the number of sound waves produced per second, while the pitch may be defined as a specific sound on a musical scale (for example a b-flat note). Galileo noted that for two sounds the ratio of their frequencies determines the perceived pitch of the sound. He also experimentally determined that it was the physical characteristics of the vibrating string, namely its mass, length, and amount of applied tension, that determined the frequency of the generated sound. A mathematical relationship between the mass and frequency was established later in the century by the French mathematician Marin Mersenne. Mersenne also introduced the idea of **harmonics**, which are vibrations that occur at ratios (2x, 3x, ½x, etc.) of the original frequency. His mathematical principles form the basis of many modern studies of music. By the end of the century Robert Hooke applied this knowledge to construct a device that generated a sound of a specific frequency.

Galileo was also one of the first in this century to demonstrate that sound travels at a specific speed. In his experiment Galileo fired a gun from a mountaintop, and then measured the time between when the flash of the shot was observed and the sound of the gun was heard. This experiment fostered an interest in determining a value for the speed of sound and as such a number of scientists duplicated Galileo's method. The French mathematician Pierre Gassendi was one of the first to employ Galileo's method. While his calculated value was a little high, since he obtained a value of 478.4 meters per second (1,569.6 feet per second), he did experimentally determine that frequency of sound does not influence the speed. More detailed experiments were conducted by the Italian scientists Vincenzo Viviani, an assistant of Galileo's, and Giovanni Borelli (1608–1679). Using the same experimental procedure they arrived at a value of 350 meters per second, which given the accuracy of scientific instruments during this time is a fairly close approximation to the modern value of 331.29 meters per second (established in 1986). The 17th-century value is even more remarkable in that the researchers did not recognize that temperature and humidity influence the speed of sound, and thus did not compensate for them in their calculations.

Perhaps the greatest advancement in the study of sound during the 17th century was the realization that sound behaved in a wave-like manner. While Aristotle had suggested that the movement of sound was a property of air, many during the 17th century believed that when sound is generated undetectable particles left the source to be detected by the ear. This idea was proved to be incorrect with the invention of the vacuum pump (see VACUUM PUMP). A number of scientists had proposed that sound behaved like waves on a pond. However, the experimental evidence for this was lacking. The inventor of the vacuum pump, the German scientist Otto von Guericke, made some preliminary studies in which he demonstrated that sound did not travel in a container from which the air had been removed. Similar experiments were performed by the German scientist Athanasius Kircher. Unfortunately, the observations that these men had made were not entirely correct. Early vacuum pumps lacked the ability to extract the air completely from the container, and unless there is a perfect vacuum, sound may be transmitted. As noted by several physicists, what Guericke and Kircher had observed was a difference in the density of the air and the container walls that prevented the sound from being transmitted. However, these experiments did attract the interest of Robert Boyle, who had displayed an intense interest in the properties of gases (see BOYLE'S LAW; GASES). Boyle modified the previous experiments and subsequently monitored the level of the sound as the air was evacuated from the chamber. He noted that the sound decreased in intensity as the air pressure decreased, and thus it was the air itself that was responsible for the transmission of sound. This provided the needed support for the wave theory of sound. The English scientist Isaac Newton expanded the theory by proposing that the movement of sound was actually due to the collision of atoms. Newton mathematically suggested that there was an inverse relationship between the speed of sound and the density and elasticity of the medium, a fact that was not proven experimentally until the 19th century.

While a number of advances in the study of sound occurred during the 17th century, these primarily served as a foundation for later investigations over the next two centuries. During this time more detailed analyses of the atomic nature of sound helped explain in greater detail the relationship between the frequency and pitch. In addition the wave-like nature of sound was later explored in liquids and solids. The 17th-century investigations of sound were primarily associated with the study of other topics, for example vacuums, atomic theory, and gases. Many of these experiments are still duplicated in beginning physics courses, an indication of the value of the experimental method originating in the 17th century. Sound has found a wide variety of applications in modern science. For example, very-low-frequency sound has been used to communicate with submarines under the ocean. Sound waves are frequently used in modern medicine in ultrasound analysis, where high-frequency sounds are used to produce images of internal organs. However, from a more historical perspective the studies of acoustics and the nature of sound indicated a change in scientific thinking from the purely philosophical methods of the ancient Greeks, toward a more experimental approach to examining the natural world.

Selected Bibliography

Lloyd, G. E. R. *Early Greek Science: Thales to Aristotle.* New York: W. W. Norton, 1970.
Taton, Rene, ed. *The Beginnings of Modern Science.* New York: Basic Books, 1958.

Wolf, A. *A History of Science, Technology and Philosophy in the 16th and 17th Centuries.* New York: Macmillan, 1968.

Star Atlases (1603–1677): Perhaps the earliest application of mapmaking of the night sky began with the creation and naming of the constellations. To early human civilizations the stars appeared to form recognizable patterns and shapes. Furthermore, the appearance of a specific pattern often coincided with important events on Earth. This was probably most relevant to agriculture where the arrival of a constellation would indicate the time for planting or harvesting of the crops. If we consider the naming of the constellations to be the oldest instances of star map creation, then the question may be raised as to when the first of these maps was constructed. The exact ages of some constellation names are difficult to determine but some experts suggest that some of the larger ones, Ursa Major for example, may have originated over 15,000 years ago. From a more modern perspective, catalogues of the stars had been prepared as early as the ancient Egyptians and Mesopotamians (ca. 700 B.C.E.), although there were probably earlier attempts that did not survive until the present age. This process would continue with the ancient Greeks. The Greeks had devised a coordinate system for the night sky that listed their longitude and latitude in relation to the sun's apparent path, or ecliptic. The astronomer Hipparchus used this practical system, which was also in use by the Chinese, to catalogue the positions of the stars around 120 B.C.E. Two centuries later, Ptolemy would publish a two-volume star atlas as part of his epic work entitled the *Almagest*. *Almagest* would include the positions of over 1,000 stars, as well as other objects like nebula (see NEBULA). After the destruction of the Great Library in Alexandria most of the early work on star maps was lost to Western cultures. However, portions of Ptolemy's work would be preserved in the Middle East by the Arabs, who would also make improvements on the positions and magnitudes, or brightness, of many of the listings. While Western cultures were in the Dark Ages, mapping of the night skies would continue in China. The Chinese astronomers were significantly ahead of their Western counterparts and by the 4th century had constructed detailed catalogues of the night sky. By 940 the Chinese had developed a method of mapping the spherical night sky on a flat surface. This technique would later be called a Mercator projection, after the Belgian cartographer Gerhard Mercator (1512–1594), who developed it around 1596.

Just prior to the 17th century an interest in star map and atlas construction was once again developing. The Danish astronomer Tycho Brahe, most widely known for his revisions of the Ptolemaic system (see HELIOCENTRISM), conducted extensive observations of the night sky. In 1598 he published a partial list of his observations as an atlas of over 700 stars. However, Tycho's observations would not be confined to stars. He accumulated a wealth of information on the planets, specifically Mars. These would form the basis of Johannes Kelper's laws of planetary motion (see KEPLER'S THEORY OF PLANETARY MOTION). In 1627 Kepler published a series of tables on the planets that were based on Tycho's observations.

There would be several important developments in the construction of star atlases and maps during the 17th century. In 1603 Johann Bayer (1572–1625), a German astronomer, published a star atlas entitled *Uranometria*. Bayer updated the existing forty-eight constellations from Ptolemy's time and added an additional twelve constellations from the Southern Hemisphere. More importantly, Bayer introduced a system of nomenclature for star atlases that is still in use to this day. Of the several thousand stars that are visible to the naked eye in the night sky, only a few hundred have common names that are regularly used by astronomers. Many of these names are Arabic and indicate the importance of Arabic science in preserving Greek astronomy. Under Bayer's system each star in a constellation was assigned a Greek letter. The brightest star was assigned the letter alpha, the second brightest would be beta, and so on. Nonstellar objects, such as nebulae, were given a separate numbering system to distinguish them from stars. In the case of constellations that have more than twenty-four stars, the number of letters in the Greek alphabet, the progression continues with Latin letters. Over 1,300 stars have now been designated using Bayer's system.

While maps of the Northern Hemisphere constellations had been around for centuries, data on the Southern Hemisphere were lacking. While bright objects such as the Southern Cross and the Magellanic Clouds, which are actually nearby galaxies, had been recorded by sailors navigating below the equator, star maps of the southern skies were virtually nonexistent. The first twelve constellations of the Southern Hemisphere were named around 1595 by Pieter Keyser, a Dutch explorer. In 1677 Edmund Halley, best known for his later work on comets (see HALLEY'S COMET), would begin his career as an astronomer by spending two years at St. Helena charting the positions of over 340 Southern Hemisphere stars. A complete construction of the southern sky would not be accomplished until 1750 when the French astronomer Nicolas Louis de Lacaille (1713–1762) provided the names for fourteen additional southern constellations.

The invention of the telescope early in the 17th century sparked a renewed interest in astronomical observations. The invention of the telescope promoted the invention of instruments designed to enhance astronomical observations. The micrometer, a device used to measure distances using a telescope (see MICROMETER), and the pendulum clock (see PENDULUM CLOCK) gave astronomers an increased accuracy in their observations. This would lead to the development of more detailed star atlases. Primary among these would be the *Uranographia*, published in 1687 by the Polish astronomer Johannes Hevelius. In this work Hevelius would not only chart the locations of over 1,500 stars, but would also propose the addition of seven new constellations.

The refinement of star atlases would continue into the 18th century. As the first Royal Astronomer, the Englishman John Flamsteed (1646–1719) constructed the most extensive catalogue of visible stars to date. In *Atlas Coelestis*, published in 1729, Flamsteed gave the positions of over 3,000 stars. He also provided an alternative system of nomenclature to the Greek system developed by Bayer. Flamsteed numbered the stars in the constellation by their right **ascension**, a method of measuring astronomical location based on the objects' east-west coordinates. Furthermore, the development of atlases would not be limited to Europe. By 1757 the Chinese had composed an atlas of similar size using technologies they obtained from Europe. Later in the century James Bradley, an Englishman, compiled a catalogue of 3,222 stars. This highly accurate work would assist astronomers and physicists in their analysis of stellar motion well into the next century.

By 1930 the International Astronomical Union had established a permanent list of eighty-eight constellations. Updated atlases continue to be published as technological improvements increase the ability of astronomers to detect fainter objects. Today, extensive star atlases and catalogues exist for almost any class of stellar object. Many of these are based on the observations of the 17th and 18th century and follow the nomenclature principles set forth by Bayer and Flamsteed. For example, the French astronomer Charles Messier's catalogue of nonstellar objects like galaxies and nebulae is still in use today (see NEBULA).

Lists of objects based on the wavelengths of energy that they emit, such as photographic and infrared, are constantly being updated due to information being obtained from satellite-based detection systems.

The 17th century may be considered the beginning of modern astronomy. The invention of the telescope, and its subsequent improvements, during this century resulted in a rapid expansion in the amount of information available to astronomers. With this increase in data arose a need to organize the information in a format that could be easily understood in the astronomical community. The star atlases and catalogues of the 17th century reflect this. While many of these catalogues may seem primitive in comparison to the massive databases available to modern astronomers, their organization and principles set the foundation for centuries to follow. To this day, these atlases provide a record of observations that aid modern scientists in their calculation of stellar motion and astrophysics.

Selected Bibliography

Evans, James. *The History and Practice of Ancient Astronomy*. New York: Oxford University Press, 1998.

Gurshtein, Alexander. "In Search of the First Constellations." *Sky & Telescope* 93, no. 6 (June 1997).

Motz, Lloyd, and Jefferson H. Weaver. *The Story of Astronomy*. New York: Plenum Press, 1995.

Ridpath, Ian. *Norton's Star Atlas and Reference Handbook*. 19th ed. Essex, UK: Addison Wesley Longman, 1998.

Wilson, Robert. *Astronomy through the Ages: The Story of the Human Attempt to Understand the Universe*. Princeton, NJ: Princeton University Press, 1997.

Steam Engine (1601–1698): The use of pressurized steam to power mechanical devices is not a recent technological development. As far back as the 1st century simple steam-powered machines were in development by the ancient Greek mathematician Hero of Alexandria (ca. 75). Hero is well known for his studies of air-driven devices and the belief that all forces of nature could be explained by mechanical principles. He is also recognized as being the first to develop a machine powered by steam, called an aeoliphile. The aeolipile was little more than a vacant sphere with several bent tubes protruding from it. It was mounted on a hollow tube that delivered steam from a cauldron to an inside cavity. As the pressure inside of the cavity increased, steam escaped from the bent tubes causing the device to spin. Although this is considered to be the first device that converted the energy of the steam into motion, it attracted little attention from the scientific community and instead was frequently used to entertain audiences. For the next fifteen centuries little progress was made on the development of steam-powered machines. A few instruments were proposed, such as a steam-powered gun designed by the renowned Italian inventor Leonardo da Vinci. However, few of these machines were actually constructed and none of them apparently had any significant impact on the science or progress of the times.

In the 17th century there was a renewed interest in the invention of steam-powered machines. As with many of the discoveries of this time, there were multiple inventors working on the problem from a wide variety of perspectives. Some of the earliest uses of steam power were targeted toward the creation of a vacuum and the movement of water. In the opening years of the 17th century the Italian alchemist Giambattista della Porta (1535–1615) proposed a machine in which the force of steam could be used to raise a column of water. The machine functioned in a manner such that as the steam condensed it created a vacuum. The surrounding water then rushed in to fill the vacuum. This is considered to be one of the earliest forms of a pumping device. However, it differed from Hero's invention in that it utilized steam to do the work, whereas the aeoliphile relied on compressed air. While della Porta's machine does not appear to ever have been used industrially, the concept of using steam to move water did gather considerable attention.

The English inventor Edward Somerset (1601–1677) designed the first steam-powered machine that moved beyond the purely experimental stage and was actually put to work. Around 1663 Somerset invented what he called a water-commanding device. This machine functioned in much the same manner as della Porta's in that it used steam to create a vacuum to move water. By using a high-pressure boiler that provided steam to two condenser tanks, the instrument was capable of raising water to a height of approximately forty feet. The dual condenser tanks increased the efficiency of the machine in that they allowed the operator to alternate between the tanks, emptying one while the other was collecting steam. In 1663, Somerset obtained a ninety-nine-year patent from the English Parliament for this invention.

The next substantial development in steam-powered machines was made by the French-English physicist Denis Papin (1647–ca. 1712). Papin was an associate of the Dutch scientist Christiaan Huygens. In addition he worked closely with the English chemist Robert Boyle. In 1681 Papin reported that he had invented a pressure cooker, which he called a "digester," that used the power of pressurized steam to break down bone material. As a precursor of the modern pressure cooker, this instrument was important since it was the first steam-driven instrument to utilize a safety valve to relieve excess pressure in the tank. A second improvement to steam engines that Papin introduced was the concept of using a piston to transfer power (see HYDROME-CHANICS). Unfortunately, Papin's initial attempts to design a steam-powered piston were unsuccessful. However, as an assistant to Huygens, Papin had access to a previous experiment in which Huygens used an explosive charge to compress gas. The compression of the gas in Huygens's engine was sufficient to move a piston. By modifying this device to use steam, Papin was able to develop enough of a vacuum to move the piston in his own machine. Papin applied this method of generating pneumatic power to a number of inventions including a pump to remove water from mines (1695) and a steam-powered model boat (1705). The removal of water from mine shafts was a significant problem for the mining industries and attempts had been made by scientists earlier in the century to address the problem (see BOYLE'S LAW; VACUUM PUMP). While the vacuum pump held some promise for the removal of water, until the invention of the steam engine it lacked the power necessary for moving large amounts of water efficiently. Papin's designs served as the inspiration for several of the 18th-century steam-driven machines.

Another inventor of the 17th century who worked with steam engines was the Englishman Thomas Savery (ca. 1650–1715). Savery was also attracted to the concept of using steam-powered machines to clear excess water from mine shafts. Savery's machine employed two chambers, a boiler and a condenser. The configuration of these chambers allowed for the continuous processing of water into the boiler, which meant that the device did not need to be drained periodically. While he did not use safety valves, an obvious hazard for a steam-driven apparatus, he did introduce the use of a gauge to measure the height of water within the device. Savery's inventions were a definite improvement over the horse-drawn method of removing water from mine shafts, but they were not without their limitations. The pumps had difficulty handling very high steam pressures and did not have the ability to raise water more than thirty or forty feet (see BOYLE'S LAW). This could be corrected by using the pumps in tandem, with steam pumps on multiple levels in mine shafts, to move the water to the surface. Furthermore, some found application in the delivery of water to tall buildings. Later in the 18th century improvements in the design and capacity of these machines made them more practical and efficient for the mining industry.

Over the course of the 17th century a number of improvements were made to steam-driven instruments. Steam engines started the century as crude devices, many of which existed as no more than abstract theoretical concepts. However, by the end of the century the inventions of Papin and Savery were beginning to provide benefits for society at large. This next generation of machines, with their improved methods of controlling and delivering steam power, started to address some of the needs of industry, specifically those with mining interests. This trend would continue well into the next two centuries as steam power became one of the primary forces of locomotion and industry. Steam power continues to play an important role in modern society where it is the basis for many forms of electrical power generation. An example is the method of power generation at nuclear and fossil fuel plants. In some areas of the world, for example Iceland, the hydrothermal power of steam is used to heat homes directly. These are all due in part to the innovation of 17th-century inventors such as della Porta, Papin, and Savery.

Selected Bibliography

Taton, Rene, ed. *The Beginnings of Modern Science.* New York: Basic Books, 1958.
Wolf, A. *A History of Science, Technology and Philosophy in the 16th and 17th Centuries.* New York: Macmillan, 1968.

T

Telescope (1608–1672): Well before the time of the ancient Greeks, early civilizations devoted considerable time to the construction of temples to help predict and explain the patterns in the night sky. Archeological investigations in our century have gathered considerable evidence that Stonehenge, the Egyptian pyramids, as well as Indian and Mayan temples not only may have served as religious centers for their times, but also may have been designed as astronomical observatories. While these structures appear to have had some accuracy in the prediction of events such as eclipses and the passage of seasons, they were limited to observations that could be made only with the naked eye.

Even the ancient Greeks, with their philosophical interest in explaining the laws of the world around them, would so be confined in their observations. While notable Greek astronomers such as Hipparchus made significant contributions to calculations on the distance to the moon, the inability of the Greeks to visualize the true complexity of the universe would result in an oversimplification of the structure of the universe. Claudius Ptolemy, in part using data derived by Hipparchus, constructed a model of the universe in which the Earth was the central object. A series of transparent spheres surrounded the Earth, on which the planets, moon, and stars rode. This spherical nature of the heavens, called the Ptolemaic System, would inaccurately guide both scientific investigations and religious beliefs until the invention of the telescope some sixteen centuries later (see HELIOCENTRISM).

To this day there remains considerable uncertainty on the early history of the telescope. While the use of the term *telescope* did not occur until the 17th century, investigations on the properties of light, specifically the ability of glass to bend and magnify light beams, dates back several centuries earlier (see OPTICS). Perhaps one of the first written descriptions of the uses of lenses to view objects at a distance was by the English philosopher Roger Bacon. Often credited with the invention of the first eyeglasses, Bacon conducted many experiments investigating the properties of light. In his book entitled *Opus Majus*, Bacon gave a description of how two lenses, when configured properly, would enable the viewer to see objects at a distance more clearly. While what Bacon had described was a primitive telescope, Bacon did not propose that the phenomena had scientific application, but rather he appeared to use the observation as data to support his developing philosophy on the physical characteristics of light. Several centuries later, Giambattista della Porta conducted similar investigations on the properties of light. In his lifetime della Porta published several descriptions of how convex and concave lenses could be combined to see objects at a distance. Unfortunately, the majority of his discoveries were published as descriptions of magic, and not science. Therefore, it is possible that these were the first examples of a functioning telescope, or one that was designed specifically to view objects at a distance, but the format in which they were printed probably means that these findings had very little influence on the scientific community.

The phenomena that both Bacon and della Porta were investigating was the ability of lenses to refract light. Unlike reflection, in which the path of light is redirected, refraction is the bending of the path of light (see Figure 23). When done correctly, as with the examples of convex and concave lenses, the effect can be a focusing of light beams at a given location. Since light scatters as one moves farther from an object, the lenses serve to focus the light and produce a clear image. The problem with the lenses constructed during the time of both Bacon and della Porta was that they often had significant imperfections in their smoothness. This resulted in a lack of clarity in the image that made application of the principles difficult. However, by the beginning of the 17th century the technology in lens production was beginning to improve. And as this technology improved, so would the quality of the telescopes being produced. This is reflected in the fact that during the 17th century telescope design went through three distinct phases, each associated with an improvement in the optical design of the instrument.

Refracting Telescopes

Credit for the first true telescope, or one in which the lenses were mounted in a permanent structure, is most often given to Hans Lip-

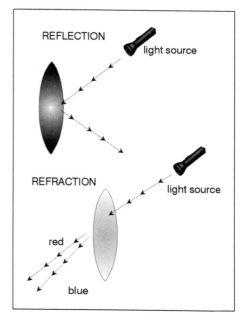

Figure 23. The principles of reflection and refraction of a light beam by a lens. In the lower figure, the refraction of the light has caused the beam to be split into red and blue wavelengths.

pershey, a Dutch eyeglass manufacturer. In 1608, while working on the use of different lens combinations to enhance magnification, workers in Lippershey's shop discovered, most likely by accident, that certain lens combinations would make distant objects appear closer. This was the same phenomenon that had previously been reported by Bacon and della Porta. Later, others such as Zacharias Janssen claimed to have made this discovery before Lippershey's group. Some accounts even suggest that it was Lippershey's children who made the discovery. Regardless, unlike his predecessors and competitors, Lippershey realized the potential that this discovery represented. To make the instrument more practical, Lippershey constructed a metal tube to house the lenses in the proper configuration. What Lippershey had constructed was the first telescope, although they were called "lookers" or "optic tubes" until a few years later when the term *telescope* was adopted. Lippershey's invention saw immediate application in the naval wars between Spain and the Netherlands as a means of determining the strength of the enemy at a distance. Despite the secrecy surrounding the first telescopes, news of the instrument was soon widely available in Europe, and many, such as Galileo, set forth to duplicate Lippershey's work. While the discovery of the telescope may appear to be a simultaneous discovery by a number of investigators, it was the development of the observations and its application to the needs of society that resulted in Lippershey's place in history. Thus, the invention of the instrument that would probably have the greatest impact on the scientific achievements of the 17th century has its origins as a weapon of war.

By some accounts, Galileo is credited with the invention of the telescope. While it is clear that Galileo did not invent the telescope before Lippershey, he is credited with converting the instrument from maritime to scientific use. Galileo was not interested in the physics of how a telescope functioned, instead he realized the potential of the instrument to explore the structure of the heavens. When Galileo first looked through the telescope, his observations immediately challenged the prevailing Ptolemaic concept of an earth-centered universe. One of Galileo's first discoveries was that the moon had physical characteristics similar to those found on Earth, with mountains and valleys and what appeared to be seas (see THE MOON). Furthermore, when Galileo focused his telescope on the Milky Way, he observed an ocean of previously undetected stars. These observations were in direct conflict with the Greek philosopher Aristotle's views on the structure of the universe, and this caused Galileo to begin to question the existing model of the structure of the heavens.

Perhaps the greatest discovery to be made in the 17th century with the use of a telescope would have occurred when Galileo first turned his attention to the planets. The movement of the planets in the heavens had consistently challenged the Ptolemaic system. In the previous century, Tycho Brahe (Tycho) had developed elaborate models on the movement of the planets and in doing so had combined the Ptolemaic and Copernican models of the solar system (see HELIOCENTRISM). He suggested that while the planets orbit the sun in a Copernican method, the sun itself orbited the Earth. When Galileo examined the planets through his telescope he saw evidence suggesting a much different scenario. When examining Jupiter, Galileo detected the presence of four satellites. Subsequent observations of the satellites indicated that these objects appeared to be in orbit around Jupiter (see JUPITER). Galileo concluded that if moons were in orbit around Jupiter, then it was equally possible that the Earth and the other planets were in orbit around the sun. Additional studies of the other planets (see MARS; SATURN) convinced Galileo of the validity of the Copernican model (see HELIOCENTRISM). In a period of less than a decade from the time it was first invented, the telescope had provided evidence that challenged theories of the universe that had persisted for over 2,000 years.

However useful the telescope was to astronomical observations of the early 17th century, the instrument did have its limitations. In refraction, the differences in the wavelengths of light (see OPTICS) do not cause the light beam to separate when passed through glass (see Figure 23). Violet light, with its short wavelength, is refracted the most while the long wavelengths of the red portion of the spectrum are refracted the least. This unequal refraction of the light beam resulted in formation of a colored halo surrounding light images, called a chromatic aberration. Some corrections were possible by lengthening the distance between the lenses. However, this tended to make the instruments difficult to handle. While advances in lens technology in later centuries would improve the capabilities of the refracting telescope, improvements to the telescope in the later half of the century would concentrate on focusing an intact light beam to the viewer by reducing refraction. The next two phases of 17th-century telescope design would present major improvements to the Galilean telescope in an attempt to solve the focusing problems of the refracting telescopes.

The invention of the refracting telescope had a significant impact on the era of scientific exploration that was to begin in the 17th century. As one of the first instruments designed specifically for scientific investigations, the invention of the telescope would force mankind to address its position in the universe. No longer would the old teachings of an Earth-centered universe prevail. The discoveries by Galileo and others would indicate that the universe was far more complicated than originally thought.

Long Telescopes

The refracting telescopes of the early 17th century, while primitive in design, had opened up the universe for a new generation of astronomers. Investigations of the complex nature of the Milky Way, the discovery of nebulae (see NEBULA), and insight into the true origin of comets (see HALLEY'S COMET) were all made possible by the early refracting telescopes. However, the early telescopes, such as those used by Galileo, were initially designed for terrestrial observations. These refracting, sometimes called Galilean, telescopes magnified distant objects by passing the light through a convex lens. As the light passed through the lens, the path of the light was bent, or refracted, to a focal point (see Figure 23). The lenses of the early telescopes rarely, however, focused the light on a single point. Thus, the images viewed through these telescopes often appeared fuzzy. Since these problems are related to the shape of the lens, they are called spherical aberra-

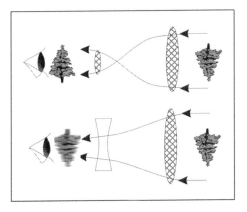

Figure 24. The differences in the refraction of light by a refracting telescope (bottom image) and the long telescopes designed by Kepler (top image). Note that the configuration of the lens in the long telescope results in an inverted image.

tions. Another major disadvantage of these instruments was that different wavelengths of light are not refracted equally. As the light passes through the convex lens, the violet end of the spectrum has a greater refraction than that of the red wavelengths (see Figure 23). Because of this, the blues are separated from the spectrum, which results in a colored halo surrounding the images. This phenomenon is called a chromatic aberration.

The invention of the telescope early in the 17th century developed a scientific interest in the physical properties of light, or optics (see OPTICS). One result of these studies was a greater understanding of the process by which light is bent as it passes through a prism or piece of glass. The German astronomer Johannes Kepler applied his study of optics to define the causes of spherical aberrations common to early telescopes. Kepler studied the principles of refraction in convex lenses and suggested the design of a new type of optical lens that would reduce the severity of the spherical aberrations. Kepler's theories would later be supported by mathematicians, such as Willebrord Snell and René Descartes, who had also developed an interest in the study of optics. Unfortunately for Kepler, the technology did not exist at the time for the creation of these new complicated lenses. As an intermediate step, Kepler proposed that the length of the telescope be increased. By doing so, the light beams would have the opportunity to converge at a single point. Unlike a Galilean telescope, whose short length often forced the eyepiece to be placed before the focal point of the image, the length of the Keplerian telescope would allow the focal point to occur before the eyepiece. As the light passed through the focal point, it would begin to diverge again slightly (see Figure 24). In a Keplerian telescope the eyepiece was placed after the light beam diverged from the focal point. Since the light was focused on a single point prior to diverging, the result was a clearer image. Unlike a Galilean telescope, the view through the eyepiece of this new instrument would be inverted. The minor disadvantage that the image was inverted was outweighed by the increase in clarity that these instruments offered. It was possible to add additional lenses to correct the inverted image. However, each additional lens reduced the quantity of light received from the source. Versions of Keplerian telescopes with corrected lenses found use in society as transits for terrestrial mapping and the telescopic gun sight. Thus, in comparison to a Galilean telescope, which allowed both astronomical and terrestrial viewing, this new generation of telescopes would be the first telescope designed specifically for astronomers. With their ability to focus more light, these longer Keplerian telescopes quickly replaced the Galilean version.

As noted, the invention of the telescope pioneered the study of optics. The 17th century saw mathematicians, such as Descartes, attempt to explain how light is refracted as it passes though a lens. Their work suggested that by decreasing the angle of refraction, more of the light could be focused on a single point, decreasing spherical aberrations and increasing the clarity of the image. This was accomplished by altering the shape of the lenses from the convex format of the Galilean telescopes to lenses that were less spherical, or paraboloid in shape. And since these lenses bent the light beam at a much smaller angle, the light beam would separate less, which in turn would reduce chromatic aberrations. Unfortunately, since the light beam was bent at a smaller angle, the length of the telescope had to be increased so as to allow enough distance after the lens to focus the light on a single point. The lesser the angle of refraction, the greater was the need to increase the length of the telescope. When added to the improvements implemented by Kepler, the result was the formation of an instrument called the long telescope. Once invented, these instruments, with reduced chromatic and spherical aberrations, became a favorite among the astronomers of the 17th century.

The long telescopes were the next major phase in the evolution of the telescope. These longer instruments possessed higher magnification potentials than the older Galilean refracting telescopes. And as the astronomers of the mid-17th century gazed through them, they observed a universe that was much more complicated than even Galileo had suggested. The first long telescopes were probably constructed sometime between 1615 and 1617. One of the first long telescope astronomers was Johannes Hevelius. Hevelius focused his studies on the moon and the planets. In his observations of the moon, Hevelius constructed the first lunar map, and in doing so he described mountain ranges and vast flat areas that he called seas (see THE MOON). The success of Hevelius with the new telescope, and his subsequent publications of his results in 1647, inspired other astronomers to adopt the new instrument.

One of these was the Dutch astronomer Christiaan Huygens. As an astronomer, Huygens is best known for his examination of Saturn. Huygens had developed a method of refining the lens-grinding process, which resulted in lenses that focused more of the incoming light at the focal point, thus increasing magnification. Using this new technology, Huygens, in 1656, announced the presence of a satellite around Saturn. Satellites had been detected previously around Jupiter (see JUPITER). However, the detection of this satellite meant that the phenomenon was relatively common in the solar system. This was yet another blow to the idea of a universe centered on the Earth. Furthermore, in earlier observations of the planets, the shape of Saturn had become somewhat of an oddity. Unlike the circular appearance of the other planets, Saturn appeared to have protrusions on both sides. By examining Saturn through a long telescope over several years, Huygens was able to determine that the protrusion was actually a flat ring that orbited, but did not touch, the planet. The explanation of Saturn's rings would take centuries to resolve, and remain a fascination for amateur astronomers to this day (see SATURN).

Other discoveries using the long telescope would follow. The Italian astronomer Giovanni Cassini focused his attention on the outer planets. In 1664, Cassini and Robert Hooke independently identified a spot on the face of Jupiter. This spot, originally called Hooke's Spot, is now called the Great Red Spot (see JUPITER). Based on the time frame of Cassini's observations, this spot appears to be a three-centuries-old storm, the details of which we are just starting to understand with the recent visit of the Galileo space probe. Cassini also was one of the first to suggest that the planets were not circular in shape, as had been suggested as far back as the ancient Greeks, but that rather they had an ellipsoidal structure. In addition, Cassini, using a long telescope that was over 136 feet long, was responsible for the identification of four additional satellites in orbit around Saturn (see SATURN). He also worked to clarify the mystery of Saturn's rings and in 1675 discovered that the ring was actually two distinct structures. The space that separates these rings is now called Cassini's Division in his honor.

The long telescope without doubt changed the way astronomers viewed the universe, and thus had a tremendous impact on the science of the 17th century. In a few short decades, our solar system was revealed to be more complicated than just a pattern of lights in the night sky. New satellites had been discovered and our companion planets turned out to be mysterious and complex. However, the long telescopes were not without disadvantages. The sheer size of these instruments, routinely over 150 feet long, made them difficult to manipulate. Furthermore, observations with these instruments were still hindered by relatively crude lenses. Although lens-grinding techniques had improved by the end of the century, it would not be until some time later that the technology would be available to grind thin lenses. In the meantime, another innovation was in store for the telescope. In 1668, the English physicist Isaac Newton would provide another solution to the problems of refracting telescopes. Rather then bend the light, Newton developed a telescope that reflected the light using a series of mirrors. This third step in the evolution of the telescope would provide the principles behind the massive terrestrial telescopes in use today.

Reflecting Telescopes

The first half of the 17th century had been a time of significant advances in the area of telescope technology. Changes to the telescope came quickly as scientists rapidly strove to understand the principle of how the instrument operated. As noted, early telescopes were plagued by both the separation of colors (chromatic aberrations) and focusing problems (spherical aberrations). Johannes Kepler redesigned the structure of the Galilean telescopes and made changes to the optical system to produce a clearer image, and in the process introduced the first generation of long telescopes. To further reduce the imaging problems common to the Galilean telescopes, mathematicians such as René Descartes proposed changes to the shape of the lens to reduce the angle of refraction, or amount that the beam of light was bent. These changes also dictated that the length of the instrument be increased. These new instruments produced higher magnifications with reduced effects of spherical aberrations. While these instruments contributed significantly to astronomy in the first half of the 17th century, the size of the telescopes made them inherently unstable and difficult to manipulate.

There were many theories being proposed on how to improve the image quality of these early telescopes. For example, Robert Hooke advanced the idea that the tube of the telescope be filled with a liquid, such as oil, to improve the focusing of the light. A more attractive alternative to the cumbersome long telescopes was to reflect, not refract, the light beam. Theoretically, reflected light would be delivered intact to the focal point and not separated as was common in current designs. And since it would no longer be necessary to calculate the relationship between the angle of refraction and length of the instrument, the overall size of the telescope could also be decreased to a more usable size. Credit for the concept of a reflecting telescope is often bestowed upon Isaac Newton, but actually the idea originated earlier in the 17th century. As early as 1616 the Italian Niccolo Zucchi (1586–1670) had suggested the use of a mirror in place of the convex lens in the existing design. Similarly, in 1663 the Scottish mathematician James Gregory (1638–1675) described a design for a reflecting telescope in his book *Optica Promota*. However, it would not be until Newton, and his research in optics, that the first functional reflecting telescope would be built.

There were two major problems that plagued the earlier attempts at designing a reflecting telescope. The first was a lack of understanding of the physical properties of light, specifically reflected light. While it could easily be observed that light is reflected from a shiny surface, the process of how the phenomenon worked was not well understood. The second problem was in the manufacturing of the reflective surface. Lens- and mirror-grinding techniques in the 17th century were primitive. The process was almost always performed directly by the investigator, which resulted in lenses that had unique imperfections and properties. It was exceptionally difficult to create surfaces that had uniform reflective properties. Isaac Newton was the first to address both problems, and thus be able to create a functioning reflecting telescope.

Newton had an intense interest in the physical properties of light (see OPTICS). Before settling on the idea of reflecting the light beam, Newton first attempted to describe the causes of chromatic aberrations produced by refracted light. In a series of experiments using a prism, Newton discovered that light actually consisted of distinct colors, which, when combined, resulted in visible light (see LIGHT SPECTRUM; LIGHT WAVES). He proposed that the cause of the chromatic aberrations was that the prism did not refract the light waves evenly. Newton then experimented with the idea of bringing a specific color into focus. However, when he focused the light on the red end of the spectrum, the remaining colors were no longer in focus. The use of

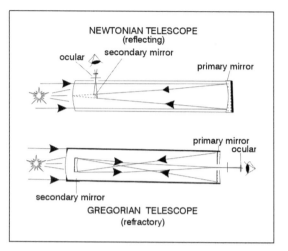

Figure 25. The differences in the optical path between a Newtonian and Gregorian reflecting telescope. Note the use of two mirrors to direct the light to the observer.

different lens-prism combinations failed to correct the problem. Finally, Newton set aside the idea of correcting the refracting telescope and directed his attention to a new design.

Newton had observed that reflected light was not subject to chromatic aberrations. In addition, he was well aware of Gregory's descriptions of reflecting telescopes, and began to design a reflective surface that would solve the imaging problems. Gregory had proposed a telescope that utilized two mirrors to direct the light beam to the observer (see Figure 25). Unfortunately, Gregory could not manufacture the necessary lenses for such a device to work. Rather than use glass, Newton decided to employ a reflective metal as his mirror. Earlier scientists had also explored the use of reflective metals but they either tarnished too easily or contained enough pigment to distort the image quality. For his mirrors, Newton used bronze, which is a combination of tin and copper, as the base metal. Newton then refined the yellow bronze surface by altering the tin-copper ratio and adding arsenic. Once formed into a concave **parabolic** shape, the result was a reflective mirror. This primary mirror would be the cornerstone of the reflecting telescope. Newton then added a second mirror to direct the light collected by the primary mirror toward the observer (see Figure 25).

In 1668, Newton constructed the first reflecting telescope. While small in stature in comparison to the massive long telescopes, the instrument had the magnification power of long telescopes much greater in size. The idea of a smaller, more manageable instrument was popular among astronomers, and in 1672 Newton presented his work to the Royal Society, an elite scientific organization in Europe. Over the next few years, other versions of reflecting telescopes were constructed as modifications were made to the path of the light entering the telescope. The Gregorian reflectors (see Figure 25), as well as later improvements such as the Cassegrain reflector, each had their own unique advantages, but still employed the same principles as the Newtonian instruments.

The invention of the reflecting telescope would have a significant impact on the field of astronomy. However, even though the reflecting telescope would be designed in the 17th century, it was not until the next century that the device would find widespread use among astronomers. The chemical nature of Newton's mirrors meant that they were still subject to tarnishing over time, and often reflected only a small percent of the available light from the source. In addition, instruments had been designed to measure stellar distances for the refractor, and these were not yet available for the new reflecting telescopes (see MICROMETER). However, without the degrading effects from chromatic and spherical aberrations, the reflecting telescopes would enable astronomers to obtain observations of the heavens with greater clarity. The minor problems with the early reflectors would be quickly worked out early in the 18th century.

The principles of reflection are still in use today in the large reflecting telescopes that sit atop the tallest mountain ranges in the world. These telescopes, and their contributions to modern astronomy and society, are the direct descendants of the reflecting telescopes first assembled by Isaac Newton in the 17th century. Although he was not the first to propose such an instrument, Newton was the first to apply his interdisciplinary understanding of scientific principles to create a functioning scientific instrument.

As a scientific instrument, the invention of the telescope may be ranked as one of the more important scientific achievements of the 17th century. In a span of less than 100 years, astronomers evolved from making simple observations of the night sky using the naked eye to being able to make detailed examinations of the heavens through a telescope. In doing so, their observations revealed a universe of a much higher complexity than originally proposed by the ancient civilizations. As a result, many of the existing theories of a geocentric universe were challenged scientifically. Furthermore, the invention of the telescope promoted the development of additional scientific instruments whose function was to enhance the accuracy of astronomical observations (see MICROMETER; PENDULUM CLOCK). This established a link between technological advances and scientific achievement that continues to this day. The invention of the telescope gave physicists the ability to collect data on the physical forces of the universe (see LIGHT SPEED; LIGHT WAVES) and in doing so enabled them to construct some of the earliest mathematical laws on the physical properties of the universe (see GRAVITATION; LAWS OF MOTION). Few other inventions in the history of science have had such a far-reaching impact on the advancement of science.

Selected Bibliography

Asimov, Isaac. *Eyes on the Universe: A History of the Telescope*. Boston: Houghton Mifflin, 1975.

Gingerich, Owen. *The Eye of Heaven: Ptolemy, Copernicus and Kepler*. New York: American Institute of Physics, 1993.

Moore, Patrick. *The Great Astronomical Revolution: 1543–1687 and the Space Age Epilogue*. Concord, MA: Albion Publishing, 1994.

Motz, Lloyd, and Jefferson H. Weaver. *The Story of Astronomy*. New York: Plenum Press, 1995.

Spangenburg, R., and D. K. Moser. *The History of Science from the Ancient Greeks to the Scientific Revolution*. New York: Facts on File, 1993.

Wilson, Robert. *Astronomy through the Ages: The Story of the Human Attempt to Understand the Universe*. Princeton, NJ: Princeton University Press, 1997.

Thermometer (1610–1702): Since early times people have estimated the temperature of an object by comparing its relative hotness or coldness to the temperature of the surrounding environment. While in the modern world this process has been automated by a number of devices, this type of measurement created obvious problems for early scientists. Without doubt many scientists of early civilizations had the need for more precise measurements and thus desired to standardize their measurements in accordance with some form of scale. Most likely there were many early experiments with the development of such devices. Notable among these was the Greek philosopher and physician Galen, who around the 2nd century proposed that the temperatures of ice and boiling water be used as reference points for a temperature scale. However, it was not until the 17th century, and the subsequent improvements in both scientific experimentation and instrumentation, that a dedicated attempt was made to quantify temperature.

A thermometer is the instrument typically associated with the measurement of heat. The term is derived from the Greek words for heat and measure. Prior to the 1620s these devices were called *thermoscopiums*, the modern term *thermometer* being first introduced around 1626. As with many of the inventions of the early scientific revolution, the invention of the first thermometer is claimed by multiple inventors and as such is a strong indication of the need for the device in the science of the times. The Italian scientist Galileo is believed to have developed a primitive temperature-measuring device around 1610, although it is possible that he experimented with such a device as early as 1592. The operation of Galileo's thermoscopium was relatively simple. A tube containing air was inverted into an open container filled with wine. A small amount of air was allowed to escape so that some wine entered up into the tube (see Figure 26). As the temperature of the air surrounding the tube increased, the volume of the air in the tube expanded. As the amount of wine inside the tube decreased, the difference could then be measured on a scale imprinted onto the tube. A decrease in air temperature had the opposite effect. Theoretically the principles of its operation were sound, but the instrument was far from accurate. First, since the liquid was exposed to the atmosphere, any changes in atmospheric pressure, such as would occur with an oncoming storm, would cause the level of the liquid to change in the same manner as a barometer (see BAROMETER). The relationship between air pressure and volume would have to wait until later in the century to be established (see BOYLE'S LAW). Second, the device must have been difficult to calibrate, as it was dependent on the volume of the tube and the physical characteristics of the wine being used. However, it is still considered one of the first attempts at thermometer design. One of the first documented uses of these instruments in scientific research is often credited to the Italian physician Sanctorius. Sanctorius made a number of important contributions to the study of medicine during the 17th century. Around 1612 he conducted experiments on the temperature of the human body using an early clinical thermometer of his own design. This was part of a long-term experiment in which Sanctorius investigated the process of human metabolism (see HUMAN PHYSIOLOGY). He used the thermometer to quantify his measurements of temperature, a significant advance in the study of medicine.

By the mid-1600s the problem of the influence of barometric pressure had been solved by the invention of liquid thermometers. The principle behind the design of this form of thermometer is relatively simple. As is the case with air, the volume of a liquid increases with a rise in temperature, while the reverse is true with a decrease in tem-

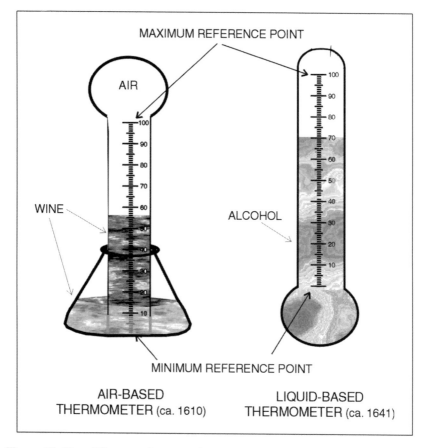

MAXIMUM REFERENCE POINT

AIR

WINE

ALCOHOL

MINIMUM REFERENCE POINT

AIR-BASED
THERMOMETER (ca. 1610)

LIQUID-BASED
THERMOMETER (ca. 1641)

Figure 26. The differences between thermometer types of the 17th century. The device to the left was an early open thermometer that was susceptible to changes in barometric pressure. The instrument on the right is a closed system and is similar in design to modern thermometers.

perature. However, unlike the exposed air thermometers, the liquid in these devices was sealed inside of a tube and thus was not susceptible to fluctuations in barometric pressure. The first of these instruments was invented around 1641 by the Grand Duke Ferdinand II (Ferdinando de' Medici, 1610–1670). To allow the liquid to expand and contract in response to temperature changes, this thermometer was designed with a small bulb at one end of the sealed tube (see Figure 26). This bulb acted as a reservoir for the liquid of choice, which in Ferdinand's case was alcohol. The alcohol may also have served to prevent the freezing of the thermometer at low temperatures. While liquids have less of a response to changes in temperature than gases, by placing a scale on the device and using small amounts of liquid it was possible to convert small changes in volume to usable temperature readings. These first liquid thermometers are often called *spirit thermometers*. While they were more accurate than the previous thermometers that used air, these early instruments did not contain fixed reference points for zero. For this reason they were considerably difficult to calibrate and standardize.

The basic principles behind the operation of the thermometer have changed little since the 17th century. The majority of the improvements that followed in the decades after its invention focused on improving the accuracy of the temperature scales. The first thermometer that utilized a fixed reference point was invented by the English scientist Robert Hooke. Hooke designed a thermometer with a single fixed reference point at zero. The scale of the thermometer was incremented in such a manner as to make readings highly accurate. Hooke's thermometer remained a standard for the Royal Society of London, a scientific organization of the 17th century, until the early part of the 18th century. This style of thermometer found application in a wide variety of sciences, including the emerging science of meteorology (see METEOROLOGY). The next major advance in calibrating the scale of the thermometer occurred at the turn of the century. In 1702 the Dutch scientist Ole Roemer designed a thermometer scale with two fixed reference points. As proposed by Galen centuries earlier, Roemer used as his references the boiling point of water and the temperature of frozen water. This became the standard for the majority of thermometers designed after this date.

Early thermometers typically used either water or alcohol as their liquid. The major disadvantage of these liquids is that they have a relatively narrow temperature range in which they exist in a liquid state. Liquid water freezes too quickly and liquid alcohol evaporates at room temperature. A solution was to find a substance that had a wider range for its liquid state. At the turn of the 18th century (ca. 1702) the French physicist Guillaume Amontons invented a device that used liquid mercury. Amontons suggested that a more reliable thermometer could be designed if one measured the change in pressure of a fixed amount of gas that was contained at a constant volume. In many ways the device was similar in operation to the thermometer designed by Galileo. However, as liquid mercury is denser than water, it helped keep the volume of air constant within the tube so that pressure could be measured. Amontons' work eventually led to the discovery of **absolute zero** for gases. This is the point at which any further lowering of the temperature does not result in a change in gas volume and all molecular motion, except resonance, ceases. The first thermometer that directly measured the contraction and expansion of mercury in response to temperature changes was designed in the 18th century by the German physicist Gabriel Fahrenheit (1686–1736). Fahrenheit played a major role in the development of thermometers during the next century.

Today there are three primary scales used by thermometers, each with its own scale and reference points. The most common scale is called the Fahrenheit scale. Named after its developer, it uses a value of 32 degrees as the reference point for the freezing of water and a temperature of 212 degrees as the reference point for boiling water. With the implementation of the metric system as a worldwide system of measurement, with the notable exception of the United States, a second temperature scale was adopted. This scale is called the Celsius scale after its inventor Anders Celsius (1701–1744). The Celsius scale divides the temperature range between the freezing and boiling points of water into 100 units, thus making calculations with the scale easier. This was initially called the Centigrade scale as it was divided into 100 units. The third scale is the Kelvin scale. This scale uses absolute zero as its fixed reference point and is used primarily in the study of physics and chemistry.

The invention of the thermometer in the 17th century had a significant impact on the science of the times. The thermometer allowed scientists to discontinue the practice of making temperature measurements in comparison to general reference points. Scientists were now able to quantify temperature data. More importantly, as the century progressed and improvements were made in the development of temperature-based scales, it was now possible for scientists to standardize and compare their measurements with those of their colleagues. The thermometer found a wide range of applications in science, from studies of meteorology to investigations of human metabolism and the properties of gases. As with many useful scientific instruments, the thermometer would continue to be improved over the next few centuries and is now an integral component of all scientific disciplines.

Selected Bibliography

Bud, Robert, and Deborah Jean Warner, eds. *Instruments of Science.* New York: Garland Publishing, 1998.

Taton, Rene, ed. *The Beginnings of Modern Science.* New York: Basic Books, 1958.

Wolf, A. *A History of Science, Technology and Philosophy in the 16th and 17th Centuries.* New York: Macmillan, 1968.

V

Vacuum Pump (ca. 1650): While sometimes called an *air pump*, in the 17th century the invention of a device to mechanically remove air from a container was actually a result of an interest in the formation of a vacuum within a tube. Pneumatic (air) devices of other sorts had been in existence for centuries. Furthermore, discussions on the nature of a vacuum were not new to 17th-century science. In the 4th century B.C.E. the Greek philosopher Aristotle proposed his four-element theory of matter (see CHEMISTRY). As a component of his theory, Aristotle stated that "nature abhors a vacuum," and proposed that a fifth compound, ether, filled the spaces between the four elements. Greek supporters of atomic theory thought that there was a void between atoms that could not be crossed (see ATOMIC THEORY) and thus could not exist in nature. These thoughts persisted until the time of the Renaissance when the development of experimental science allowed scientists to finally explore the true nature of a vacuum.

   The invention of the first mechanical device to intentionally create a vacuum in a container is credited to the German inventor Otto von Guericke. Around 1645, using this pump, Guericke was able to remove most of the air from containers to create a vacuum. He utilized two spheres to demonstrate the power of air pressure. In one experiment, Guericke connected together two fourteen-inch-diameter copper hemispheres, commonly called Magdeburg spheres after the town in which Guericke was mayor. Using one of his pumps Guericke removed the air from the interior space. As the pressure on the inside of the spheres dropped, the air pressure on the exterior held the spheres tightly together. In fact, the pressure was so strong that a team of horses was unable to separate the spheres. When the air pressure was restored, however, the hemispheres were easily separated. As this experiment was performed before the Emperor Ferdinand III, the presentation created a considerable amount of interest for the device. Thus, the development of the vacuum pump had a strong impact on future scientific investigations.

There were a number of investigations conducted in the 17th century that focused on the nature of air. Early in the century the Italian scientist Evangelista Torricelli invented a primitive barometer to measure changes in air pressure (see BAROMETER). For this device to work Torricelli was required to evacuate a portion of the air from the top of the barometric tube, although he did not identify the composition of this space. In addition, the English physicist Robert Boyle was studying changes in air pressure from a mathematical expression. He noted that changes in the pressure of a gas influence the volume of the gas. This was formulated as Boyle's Law (see BOYLE'S LAW) and had a strong influence on the development of atomic theories during the century (see ATOMIC THEORY). Boyle's work would not have been possible without the invention of the vacuum pump a decade earlier. In studies of combustion, vacuum pumps were used to determine that combustion would not occur unaided in a vacuum. This work was important in the development of early theories of combustion (see COMBUSTION). Guericke also performed an experiment (see Figure 27 for an example of his experimental system) in which he measured the weight of the spheres both before and after a vacuum was created. In doing so he was able to make some preliminary calculations as to the density of air. This contributed to the foundation of early theories of the physical properties of gases (see GASES). It was also Guericke who first used a manometer, a device that measures pressure, to detect the level of the vacuum the pump generated.

As is the case with important inventions, Guericke's simple machine was improved on during the remainder of the century. Both Robert Boyle and the English scientist Robert Hooke made significant technical improvements to the device, most of which increased the reliability of the instrument. Boyle and Hooke are also considered to be responsible for developing the pump as an instrument of science, although others such as Guericke were active in research at about the same time. By the end of the century Guericke's device, which relied on levers, had been replaced by pistons and double pumps. By reversing some of these machines it was also possible to compress gas, a useful resource in many areas of science and industry. Improvements to the vacuum pump continued throughout the 18th century with the

Figure 27. An example of the experimental system designed by Otto von Guericke for the study of a vacuum. (From *Experiments on Atmospheric Pressure*, National Library of Medicine photo collection.)

invention of a two-stage device. However, the majority of these improvements were technical; the science of the invention originated within the 17th century.

The invention of the vacuum pump in the 17th century signaled a change in the technology of science. Together with such inventions as the micrometer and barometer (see BAROMETER; MICROMETER), scientists were developing dedicated instruments to address specific scientific questions. This allowed for a more accurate examination of the natural world and the beginning of a significant revolution in science. As is often the case, this scientific invention eventually found many applications in both science and industry. Vacuum pumps were used to study the process of respiration (see GASES; HUMAN PHYSIOLOGY), combustion (see COMBUSTION), and the transmission of sound (see SOUND). In addition, using a vacuum pump powered by steam, it was finally possible to remove water from the bottom of mine shafts, a problem that had been plaguing scientists throughout the century (see BOYLE'S LAW; STEAM ENGINE). For the vacuum pump, its invention marked the end a scientific mode of thinking that had persisted since the time of the Greeks. The invention of the vacuum pump is an example of a trend in the science of the 17th century. Throughout the century scientists were developing new instruments and machines in order to examine the physical forces of nature. Many of these devices allowed scientists to overturn existing doctrines and theories. With the invention of the vacuum pump it became clear that a vacuum could be created artificially and thus was not an unnatural phenomenon.

Selected Bibliography

Ronan, Colin. *Science: Its History and Development among the World's Cultures.* New York: Facts on File, 1982.

Spangenburg, R., and D. K. Moser. *The History of Science from the Ancient Greeks to the Scientific Revolution.* New York: Facts on File, 1993.

Wolf, A. *A History of Science, Technology and Philosophy in the 16th and 17th Centuries.* New York: Macmillan, 1968.

Appendix: Entries by Scientific Field

Astronomy

Halley's Comet

Heliocentrism

Jupiter

Kepler's Theory of Planetary
    Motion

Mars

Micrometer

The Moon

Nebula

Pendulum Clock

Saturn

Star Atlases

Telescope

Biological Sciences

Botany

Botanical Classification

Cells

Plant Morphology

Microbiology

Cells

Embryology

Microscope

Physiology

Blood Circulation

Chemistry

Human Physiology

Lymphatic System

Thermometer

Zoology

Animal Generation

Cells

Embryology

Human Physiology

Chemistry

Atomic Theory

Chemistry

Gases

Phosphorus

Geology

Fossils

Geology

Hydrologic Cycle

Ocean Tides

Instrumentation

Calculating Machines

Hygrometers and Hygroscopes

Micrometer

Pendulum Clock

Slide Rule

Steam Engine

Telescope

Thermometer

Vacuum Pump

Mathematics

Analytical Geometry

Binomial Theorem

Calculating Machines

Calculus

Fermat's Last Theorem

Logarithms

Probability

Scientific Reasoning

Slide Rule

Meteorology

Barometer

Hygrometers and Hygroscopes

Meteorology

Physical Science

Atomic Theory

Barometer

Boyle's Law

Combustion

Electricity

Gases

Gravitation

Hooke's Law

Hydromechanics

Hygrometers

Kepler's Theory of Planetary Motion

Law of Falling Bodies

Laws of Motion

Light Spectrum

Light Speed

Light Waves

Magnetic Declination

Magnetism

Optics

Pendulum Clock

Sound

Vacuum Pump

GLOSSARY OF TECHNICAL TERMS

Abscissa (Mathematics). In a coordinate system, the distance of the point along the horizontal axis. The abscissa is typically called the x-axis.

Absolute Zero (Physics, Chemistry). The temperature at which all molecular motion ceases. This is a theoretical temperature, for it has never been achieved experimentally. It is approximated to occur at -273.16 degrees Celsius.

Absorption Spectrum (Physics, Biology). The spectrum of light, measured in wavelengths, that results when a pigment is placed between a light source and a spectroscope. It represents those wavelengths that are not absorbed by the pigment.

Acceleration (Physics). The change of the velocity of an object with respect to time. Mathematical analysis of acceleration was enhanced by the development of calculus around 1684.

Acoustics (Physics). The area of physical science that studies the properties of sound, including how sound is perceived.

Analytical Geometry (Mathematics). The definition of geometric shapes and curves using algebraic functions. It involves the plotting of the shape on a coordinate system to define its physical properties.

Angiosperm (Biology). A biological class of organisms that consists of members of the plant kingdom whose seeds are enclosed within specialized leaves called the carpels. They are frequently called the flowering plants, of which there are approximately 220,000 species.

Ascension (Astronomy). Also called right ascension, this is a coordinate used with declination to establish the celestial location of an object. Declination is the angle from the celestial equator while right ascension is the angle measured along the celestial sphere from the vernal equinox.

Astronomical Unit (Astronomy). Frequently abbreviated as AU, an astronomical unit represents the mean distance of the Earth from the sun. In modern astronomy an AU has been assigned a value of 149,597,870.7 kilometers or 1.581284×10^{-5} light years.

Atom (Chemistry). The smallest unit of matter that still possesses all of the chemical properties of an element.

Atomic Number (Chemistry). The number of protons in the nucleus of an element. This also refers to the element's location on the periodic table of the elements.

Atomic Weight (Chemistry). The weight of a single atom of an element. It is calculated as the sum of the protons and neutrons in the nucleus of the atom and is expressed as atomic mass units.

Atrium (Biology). In the biological sciences, the name given to a number of different cavities in the body of an animal. In humans, this term refers to the chambers of the heart that receive blood from the veins. Blood leaving the atria proceeds into the ventricles of the heart.

Axioms (Mathematics). The assumptions or truths on which a mathematical theory is based.

Biochemistry (Biology, Chemistry). A combination of the biological and chemical sciences that involves the chemical investigation of living organisms. Frequently this field focuses on the chemistry of biomolecules such as carbohydrates, proteins, fats, and nucleic acids.

Biomes (Biology). In the biological sciences this represents the largest terrestrial ecosystems. Biomes are typically studied by ecologists and are influenced by both living and nonliving factors, such as precipitation and temperature. An example of a biome is the tropical rain forest.

Biomolecules (Biology). The general classification for the group of organic (carbon-based) molecules that are used to build living cells. Proteins, carbohydrates, fats and lipids, and nucleic acids all belong to this group.

Blastoderm (Biology). In the study of embryology this is the cleavage of cells that occurs shortly after fertilization. These cells typically form the shape of a disk after division.

Botany (Biology). The branch of the biological sciences that studies the members of the plant kingdom.

Calcination (Chemistry). A process by which metal ores are heated to remove water and carbon dioxide.

Cambium (Biology). The area of living cells that exist just under the bark of a tree (as well as other members of the plant kingdom). This is where the tree grows outward by cell division (also called secondary growth) and is where the secondary transport tissues (xylem, phloem) are produced.

Centrifugal Force (Physics). A force of inertia belonging to an object in motion in a circular path. This force is directed away from the center of the curve and is the opposite of centripetal force.

Centripetal Force (Physics). For an object in a circular path, this is the force directed toward the center of the curve required by the object to continue its movement along the curved path. According to Newton's Laws of Motion, it is the opposing force to centrifugal force.

Chromatic Abberation (Astronomy, Physics). In early optical systems, the distortion of an image caused by the unequal refraction of a light wave through a lens. This results in a colored halo surrounding the image.

Chronometer (Physics). An instrument used to measure time. The chronometer was initially invented by John Harrison in the 18th century to correct stability problems associated with the use of pendulum clocks at sea. Chronometers are mounted on platforms that keep them horizontally stable.

Compression (Physics). The increase in the pressure of a liquid or gas in a closed container as the volume of the container is reduced. An example is the compression of air in the piston of an internal combustion engine.

Deformation (Physics). The change in the dimensions of an object due to the influence of mechanical stress or strain.

Density (Chemistry). The ratio of the mass of an object to its volume. It is expressed in units of mass/volume, for example, mg/ml.

Dicotyledon (Biology). Also called dicots. These are a subgroup of the flowering plants (angiosperms) that display two seed leaves (cotyledons) in early development. Dicots are also characterized as having net-veined leaves and flower parts in multiples of two or five. Approximately 165,000 species have been identified, with trees being a prime example of this group.

Differentiation (Mathematics). The process of finding the derivative of a mathematical term. A derivative is the measure of the instantaneous rate of change with respect to a variable such as time.

Diffraction (Physics). A property of an electromagnetic wave, such as light or sound. In diffraction the wave is bent around an obstacle in its path. This can be demonstrated by the ability to hear sounds originating from around the corner of a wall.

Displacement (Physics). When an object is placed into water, the displacement is the weight of the water that is replaced by the object. The amount of displacement equals the total weight of the object, including its contents. Also called the Archimedes Principle.

Echinoderm (Biology). In biological classification echinoderms are a phylum of marine organisms that display radial symmetry, such as that found in a wheel. Unlike the cnidarians (jellyfish), they possess a complex body cavity. Examples are the sea cucumber and the sea stars.

Ecliptic (Astronomy). The great circle formed by the intersection of the Earth's orbit with the celestial sphere. It is best represented by the annual path of the sun through the sky.

Elasticity (Physics). The ability of an object to return to its original configuration or shape after being exposed to some form of physical stress. In 17th-century science this is referred to as Hooke's Law.

Electroluminescence (Physics). The emitting of light after an object containing phosphorus is exposed to an electrical field.

Electromagnetic (Physics). The combination of the properties of electrical and magnetic forces, specifically electrical and magnetic fields.

Element (Chemistry). A substance that cannot be broken down into simpler substances by ordinary chemical methods. Substances that belong to the same element have the same atomic number.

Embryology (Biology). The branch of the biological sciences that studies the genetics, structure, and development of the embryos of living organisms.

Epicycles (Astronomy). In early astronomical studies, an epicycle was a small circular wobble in the orbit of a planet or moon. It was used to compensate for inaccuracies and deviations in astronomical observations and models.

Epigenesis (Biology). The theory of embryonic development that states that the structures of the embryo form gradually over time. The term was first used by William Harvey in 1651 as a counter to the theory of preformation.

Equinox (Astronomy). The point in time when the sun crosses the path of the Earth's equator. At this point night and day are of equal duration. This occurs around March 21st and September 23rd of each year.

Ether (Chemistry). Prior to the 20th century, ether was considered to be the element or medium that filled the space between all matter. The modern term defines a specific chemical compound containing oxygen. It belongs to a class of compounds called the alkoxyalkanes.

Exobiology (Biology). The area of the biological sciences that studies the possibility of life on planets other than Earth.

Frequency (Physics). The rate of repetition of a wave. If it is an electromagnetic wave, this rate is typically measured in wavelengths per second or hertz.

Function (Mathematics). The association of an object from one mathematical set to an object of another mathematical set. In mathematics, the first set is called the range and the second is labeled the domain.

Gametes (Biology). The sex cells of a living organism. In the male these consist of the sperm cells, while in the female the ovum are the gametes.

Genealogy (Biology). The study of family histories. Genealogists analyze the ancestry of a person or the genetic descent of a family or group.

Genus (Biology). In the biological sciences, a stage in the taxonomic classification of organisms. Members of a genus are further divided into smaller groups called species based on additional differences between the members.

Geocentrism (Astronomy). The belief that the Earth is the center of the universe. Any model that places the Earth as the reference point.

Geotropism (Biology). The reaction of a growing organism, usually a plant, to the influence of gravity. This serves to orient the growing organism to the gravitational field of the planet. This explains why the roots of a plant grow downward regardless of how the seed is planted in the ground.

Gymnosperm (Biology). A biological class of organisms that consists of members of the plant kingdom whose seeds are not enclosed. Also called the naked-seed plants, the most common example is the conifer. There are about 700 known species.

Harmonics (Physics). In the analysis of sound, a wavelength that is an exact multiple of another wavelength. For example, the base frequency of 20 hertz would have harmonics at 40, 60, and 80 hertz.

Heliocentrism (Astronomy). An astronomical model that centers the solar system on the sun. It is the opposite of geocentrism.

Hydraulics (Physics). A mechanical system that is operated by the action of a fluid.

Hydrodynamics (Physics). The study of the properties of a fluid as they relate to the motion of the fluid due to the application of pressure.

Hydrostatics (Physics). The study of the properties of a fluid at rest. This is usually associated with the analysis of equilibrium and pressure.

Iatrochemistry (Chemistry). The application of chemistry to the medical practices of the 16th and 17th centuries. In modern medicine this is called pharmacology.

Inclination (Astronomy). The angle between the ecliptic and the orbital plane of a planet or satellite. The deviation from the normal orbital plane of the solar system.

Inertia (Physics). The property of an object that opposes any change to the motion of the object. The inertia of an object is dependent on the object's mass.

Infinitesimals (Mathematics). Any variable that gets consistently smaller as it approaches zero as a limit. Infinitesimal analysis involves the integration of an infinite number of these values.

Integer (Mathematics). The set of whole positive and negative numbers including zero (. . . $-3, -2, -1, 0, 1, 2, 3, . . .$).

Integration (Mathematics). A function of calculus that involves finding the integral of a term or equation. The finding of an area defined by the boundaries of the graph of a given function. It is also called the antiderivative.

Invertebrate (Biology). In biological classification, the name given to all organisms that lack a vertebral column. Any organisms that do not belong to the phylum Chordata of the animal kingdom.

Isotope (Chemistry). Atoms of the same element that contain different numbers of neutrons. A neutron is an uncharged particle in the nucleus of the atom. While all isotopes of an element have the same chemical properties, some isotopes are radioactive, allowing them to be used as markers in chemical reactions.

Latitude (Astronomy). The distance of a point north or south of a reference point. In terrestrial systems, this reference point is typically the equator, while in celestial systems it is usually the ecliptic.

Librations (Astronomy). The oscillation, or wobble, of a planet or moon on its axis.

Longitude (Astronomy). The distance of a point east or west of a reference point or meridian. In terrestrial systems this is the prime meridian that runs through Greenwich, England. For celestial systems the reference is the First Point of Aries.

Luminescence (Chemistry). The emission of light from an object.

Manometer (Physics). An instrument that measures the pressure of a liquid or gas.

Mass (Chemistry). The amount of matter contained within an object. Frequently confused with weight, which is the measure of the gravitational attraction of the planet on the mass. Weight is dependent on the distance of the mass from the source.

Matter (Chemistry). Anything that has mass and occupies space.

Mean (Mathematics). The sum of a series of values divided by the number of values in the series.

Metamorphosis (Biology). The change in the body structure of a living organism over a relatively short period of time. This is frequently used to describe the change from a larval to adult stage in insects.

Microbiology (Biology). In the biological sciences that area that specializes in the study of microscopic organisms, usually smaller than 1,000 micrometers in size. Historically this has included viruses, bacteria, protistans, and fungi.

Micrometer (Astronomy). A mechanical device used in the 17th century to measure the distances between stars against the backdrop of the night sky.

Mollusk (Biology). A classification of invertebrate organisms that have a complex body cavity, a ventral muscular foot, and a tongue-like structure called a radula. Representatives of this group include the snails and clams.

Monocotyledon (Biology). Also called monocots. These are a subgroup of the flowering plants (angiosperms) that possess a single seed leaf (cotyledon) during early development. Monocots have flower parts in multiples of three or six and leaves of this group display parallel veining. Approximately 55,000 species have been identified with grasses and corn being examples of plants in this class.

Morphology (Biology). The biological study of the structure and form of living organisms.

Mutation (Biology). A change in an organism's inherited genetic characteristics. Modern science recognizes that mutations are the result of alternations in an organism's DNA.

Odometer (Physics). An instrument that measures distance traveled.

Ordinate (Mathematics). In a coordinate system, the perpendicular orientation of a point in relation to the *x*-axis.

Organic (Biology). Carbon-containing.

Oviparous (Biology). This refers to those organisms that lay eggs for reproduction.

Oxidation (Chemistry). The addition of oxygen to a compound or any reaction that results in a loss of electrons from an atom or compound. Oxidation reactions are typically coupled with reduction reactions that gain electrons.

Pangenesis (Biology). The theory that the sex cells of the body contain invisible germs that were contributed from all of the cells of the body for the purpose of passing on genetic information.

Parabolic Mirror (Physics, Mathematics). A type of mirror that produces perfectly parallel beams of light. In mathematics a parabola is the geometric shape formed by the intersection of a cone with a plane parallel to one side of the cone.

Parallax (Physics). The apparent change in the position of an object when viewed from two different vantage points. The parallax effect is greatest when the object being viewed is close to the observer.

Parenchyma (Biology). The tissues of a living organism that are composed of unspecialized cells.

Peristalsis (Biology). The sequential compression of involuntary smooth muscle tissue for the purpose of moving material through the gastrointestinal tract.

Phlogiston (Chemistry). A substance believed to be emitted from a compound during the process of combustion. Believed to be a number of substances (including hydrogen) by early chemists, the substance has been proved not to exist.

Photosynthesis (Biology). A process conducted by many autotrophic (self-feeding) organisms, and all members of the plant kingdom, that converts sunlight to chemical energy for the end purpose of forming sugar.

Pineal Gland (Biology). In higher vertebrates, an organ of the endocrine system. It is believed to be associated with diurnal (day-night) synchronization of the biological pathways.

Pistil (Biology). A female reproductive structure of a flowering plant.

Pitch (Physics). The dominant frequency of sound emitting from a source, the quality of which may be discerned from other sound sources.

Polynomial (Mathematics). An algebraic expression in the form of $a_0x^n + a_1x^{n-1} \ldots a_n$.

Preformation (Biology). The theory of embryonic development that stated the embryo was completely formed in miniature in the gametes. Some supported the idea that the ovum housed the miniature person, while others proposed that the sperm cell was the host.

Protozoans (Biology). A biological classification of unicellular organisms.

Pyrolysis (Chemistry). The breakdown of a compound by exposing it to heat.

Reflection (Physics). The return of an electromagnetic wave, such as light or sound, after striking a surface.

Refraction (Physics). The deflection of a light wave after it strikes the boundary between two different media (for example, glass and air). A refracted light passes through the media but follows a different vector than the original wave. Refraction was the basis of Snell's Law in the 17th century.

Respiration (Biology). At the organismal level this is the exchange of gases between internal tissues and the external environment. At the cellular level this term indicates the metabolic processes by which cells obtain energy from organic materials.

Retrograde Motion (Astronomy). The apparent movement of a celestial object from east to west against the backdrop of the night sky. Normal celestial motion is from west to east. It is the result of the movement of the Earth in comparison to the movement of the object.

Rotifers (Biology). A phylum of filter-feeding animals. Most members of this phylum can be viewed only with the aid of a microscope.

Sedimentation (Geology). The precipitation of particles from a water supply (ocean, river, pond, etc.) as a result of gravitational or centrifugal forces.

Septum (Biology). In anatomy, a tissue structure that separates two cavities. The septum of the heart attracted considerable attention during the 17th century.

Species (Biology). There are many definition of a species. Basically, a species is a group of interbreeding individuals who are reproductively isolated from other groups.

Spectroscope (Physics). An instrument that breaks down light into distinct wavelengths for analysis. Frequently it is used to break down the visible light spectrum (wavelengths between 400 and 700 nm).

Spherical Aberrations (Astronomy). A problem of early optical devices in which imperfections in the quality of the lens caused a blurring of the image. It was corrected using different concave-convex lens combinations and parabolic mirrors.

Stamen (Biology). The male reproductive portion of a flowering plant.

Tangent (Mathematics). A line or plane that comes in contact with the surface of a curve at a single point.

Taxonomy (Biology). The classification of living things based on any number of shared characteristics (DNA, morphology, physiology, etc.).

Theorems (Mathematics). A statement of mathematical truth that is presented along with its qualifying conditions.

Theory of Natural Selection (Biology). A theory first presented by Charles Darwin that explains the survival of the most adapted life forms and the subsequent changes in the population over time.

Thorax (Biology). The term used to describe an area of the body of a living organism. In humans the thorax is the area between the head and abdomen.

Transmutation (Chemistry). The changing of one element to another. In modern chemistry this is known to involve changes in the atomic structure, specifically the number of protons in the nucleus.

Trigonometry (Mathematics). The branch of mathematics that investigates the angles and properties of triangles.

Vacuum (Physics). An area or space that lacks any evidence of matter. Pure vacuums are rare; even the vacuum of space contains some atomic particles.

Vector (Physics, Mathematics). A quality that contains both a magnitude and direction. In mathematics the direction of the vector is indicated by an arrow while the magnitude is represented by the length of the line segment.

Velocity (Physics). The rate of change in the speed of an object over time.

Ventricles (Biology). In human anatomy this term is used to describe a cavity in the brain or heart. The ventricles of the heart receive blood from the atria and pump it under pressure to the lungs and tissues of the body.

Vortex (Physics). An intense spiral motion of matter that is confined to a given region.

Wavelength (Physics). A term used to describe the physical distance between two identical points along an electromagnetic wave.

Xylem (Biology). In members of the plant kingdom, except mosses, the tissue that transports water within the plant.

Zoology (Biology). The area of the biological sciences that focuses on the nature of the members of the animal kingdom.

SELECTED BIBLIOGRAPHY

Asimov, Isaac. *Asimov's Chronology of Science and Discovery.* New York: HarperCollins, 1994.

———. *Asimov's New Guide to Science.* New York: Basic Books, 1984.

———. *Eyes on the Universe: A History of the Telescope.* Boston: Houghton Mifflin, 1975.

Asimov, Isaac, and Jason Shulman. *Isaac Asimov's Book of Science and Nature Quotations.* New York: Weidenfeld & Nicholson, 1988.

Ball, W. W. Rouse. *A Short Account of the History of Mathematics.* Dover, DE: Dover Publishers, 1960.

Barnes-Svarney, Patricia. *The New York Public Library Science Desk Reference.* New York: Macmillan Company, 1995.

Biswas, Asit K. *History of Hydrology.* Amsterdam, Netherlands: North-Holland Publishing, 1970.

Bonelli, Fredrico, and Lucio Russo. "The Origin of Modern Astronomical Theories of the Tides: Chrisogono, de Dominis and Their Sources." *British Journal for the History of Science* 29 (1996): 385–401.

Booth, Verne H. *The Structure of Atoms.* New York: Macmillan, 1964.

Boyer, Carl B., and Uta C. Merzbach. *A History of Mathematics.* 2nd ed. New York: John Wiley & Sons, 1989.

Bradbury, S. *The Evolution of the Microscope.* New York: Pergamon Press, 1967.

Bud, Robert, and Deborah Jean Warner, eds. *Instruments of Science.* New York: Garland Publishing, 1998.

Bynum, W. F., E. J. Browne, and R. Porter, eds. *The Dictionary of the History of Science.* Princeton, NJ: Princeton University Press, 1981.

Cajorie, Florian. *A History of Physics.* New York: Dover Publications, 1962.

———. *A History of the Slide Rule and Allied Instruments.* Mendham, NJ: Astragal Press, 1994.

Centore, F. F. *Robert Hooke's Contributions to Mechanics—A Study in Seventeenth Century Natural Philosophy.* The Hague, Netherlands: Martinus Nijhoff, 1970.

Christianson, Gale E. *In the Presence of the Creator.* New York: Macmillan, 1984.

Clark, David H., and F. Richard Stephenson. *The Historical Supernovae.* Oxford, UK: Pergamon Press, 1977.

Cobb, Cathy, and Harold Goldwhite. *Creations of Fire, Chemistry's Lively History from Alchemy to the Atomic Age*. New York: Plenum Press, 1995.

Cohen, I. B. C. *Revolution in Science*. Cambridge, MA: Harvard University Press, 1985.

Cooke, Robert. *The History of Mathematics: A Brief Course*. New York: John Wiley & Sons, 1997.

Drake, Stillman. *Galileo Studies, Personality, Tradition and Revolution*. Ann Arbor: University of Michigan Press, 1970.

Duffin, Jacalyn. *History of Medicine: A Scandalously Short Introduction*. Toronto: University of Toronto Press, 1999.

Eales, Nellie B. "The History of the Lymphatic System, with Special Reference to the Hunter-Munro Controversy." *Journal of the History of Medicine and Allied Sciences* 29, no. 3 (1974): 280–294.

Emiliani, Cesare. *Dictionary of the Physical Sciences*. New York: Oxford University Press, 1989.

Evans, James. *The History and Practice of Ancient Astronomy*. New York: Oxford University Press, 1998.

Eves, Howard. *An Introduction to the History of Mathematics*. 3rd ed. New York: Holt, Rinehart & Winston, 1969.

Fagan, Brian. *From Black Land to Fifth Sun—The Science of Sacred Sites*. Reading, MA: Addison-Wesley, 1998.

Frisinger, Howard H. *The History of Meteorology: To 1800*. New York: Science History Publications, 1977.

Galilei, Galileo. *Dialogue Concerning the Two Chief World Systems*. Translated by Stillman Drake. Berkeley: University of California Press, 1953.

Gasking, Elizabeth. *Investigations into Generation: 1651–1828*. Baltimore, MD: Johns Hopkins University Press, 1967.

Gilbert, William. *On the Lodestone and Magnetic Bodies*. In *Great Books of the Western World*, vol. 28, edited by Robert M. Hutchins. Chicago: Encyclopedia Britannica, 1952.

Gingerich, Owen. *The Eye of Heaven: Ptolemy, Copernicus and Kepler*. New York: American Institute of Physics, 1993.

Gohau, Gabriel. *A History of Geology*. New Brunswick, NJ: Rutgers University Press, 1990.

Grattan-Guiness, I., ed. *Companion Encyclopedia of the History and Philosophy of the Mathematical Sciences*. London: Routledge, 1994.

Gurshtein, Alexander. "In Search of the First Constellations." *Sky & Telescope* 93, no. 6 (June 1997).

Hall, A. Rupert. *From Galileo to Newton*. New York: Dover Publications, 1981.

Harvey, William. *Anatomical Exercises on the Generation of Animals*. In *Great Books of the Western World*, vol. 28, edited by Robert M. Hutchins. Chicago: Encyclopedia Britannica, 1952.

———. *On the Circulation of Blood*. In *Great Books of the Western World*, vol. 28, edited by Robert M. Hutchins. Chicago: Encyclopedia Britannica, 1952.

———. *On the Motion of the Heart and Blood in Animals*. In *Great Books of the Western World*, vol. 28, edited by Robert M. Hutchins. Chicago: Encyclopedia Britannica, 1952.

James, Robert C. *Mathematics Dictionary*. 5th ed. New York: Van Nostrand Reinhold, 1992.

Katz, Victor J. *A History of Mathematics: An Introduction*. Reading, MA: Addison-Wesley Longman, 1998.

Keller, Eve. "Embryonic Individuals: The Rhetoric of Seventeenth-century Embryology and the Construction of Early-modern Identity." *Eighteenth-Century Studies* 33, no. 3 (2000): 321–348.

Kline, Morris. *Mathematical Thought from Ancient to Modern Times*. New York: Oxford University Press, 1972.

Kozhamthadam, Job. *The Discovery of Kepler's Laws: The Interaction of Science, Philosophy and Religion*. Notre Dame, IN: University of Notre Dame Press, 1994.

Krebs, Robert E. *The History and Use of Our Earth's Elements: A Reference Guide*. Westport, CT: Greenwood Press, 1998.

———. *Scientific Development and Misconceptions through the Ages: A Reference Guide*. Westport, CT: Greenwood Press, 1999.

———. *Scientific Laws, Principles and Theories: A Reference Guide*. Westport, CT: Greenwood Press, 2001.

Lloyd, G. E. R. *Early Greek Science: Thales to Aristotle*. New York: W. W. Norton, 1970.

Maistrov, L. E. *Probability Theory—A Historical Sketch*. New York: Academic Press, 1974.

Michel, Henry. *Scientific Instruments in Art and History*. New York: Viking Press, 1967.

Middleton, William E. *The History of the Barometer*. Baltimore, MD: Johns Hopkins University Press, 1964.

Moore, Patrick. *The Great Astronomical Revolution: 1543–1687 and the Space Age Epilogue*. Concord, MA: Albion Publishing, 1994.

Morton, A. G. *History of Botanical Science*. London, UK: Academic Press, 1981.

Motz, Lloyd, and Jefferson H. Weaver. *The Story of Astronomy*. New York: Plenum Press, 1995.

Needham, Joseph. *A History of Embryology*. New York: Abelard-Shuman, 1959.

Newton, Isaac. *The Principia*. Translated by J. Bernard Cohen and Anne Whitman. Berkeley: University of California Press, 1999.

Park, David. *The Fire within the Eye: A Historical Essay on the Nature and Meaning of Light*. Princeton, NJ: Princeton University Press, 1997.

Parkinson, Claire L. *Breakthroughs: A Chronology of Great Achievements in Science and Mathematics, 1200–1930*. Boston: G. K. Hall & Co., 1985.

Pinto-Correia, Clara. *The Ovary of Eve: Egg and Sperm and Preformation*. Chicago: University of Chicago Press, 1997.

Ptolemy, C. *The Almagest*. In *Great Books of the Western World*, vol. 16, edited by Robert M. Hutchins. Chicago: Encyclopedia Britannica, 1952.

Pullman, Bernard. *The Atom in the History of Human Thought*. New York: Oxford University Press, 1998.

Rennie, John, ed. "Key Space Explorations of the Next Decade. The Future of Space Exploration." *Scientific American* 10, no. 1 (1999).

Ridpath, Ian. *Norton's Star Atlas and Reference Handbook*. 19th ed. Essex, UK: Addison Wesley Longman, 1998.

Ronan, Colin. *The Atlas of Scientific Discovery*. London: Quill Publishing Ltd., 1983.

———. *Science: Its History and Development among the World's Cultures*. New York: Facts on File, 1982.

Sagan, Carl, and Ann Druyan. *Comet*. New York: Random House, 1985.

Salzberg, Hugh W. *From Caveman to Chemist—Circumstances and Achievements*. Washington, DC: American Chemical Society, 1991.

Scott, Joseph F. *The Scientific Work of René Descartes (1596–1650)*. London: Taylor & Francis, 1976.

Serafini, Anthony. *The Epic History of Biology*. New York: Plenum Press, 1993.

Sheehan, William. *The Planet Mars: A History of Observation and Discovery*. Tucson: University of Arizona Press, 1996.

Smith, Ray F., Thomas E. Mittler, and Carroll N. Smith, eds. *History of Entomology*. Palo Alto, CA: Annual Reviews, 1973.

Spangenburg, R., and D. K. Moser. *The History of Science from the Ancient Greeks to the Scientific Revolution*. New York: Facts on File, 1993.

Stillwell, John. *Mathematics and Its History*. New York: Springer-Verlag, 1989.

Swetz, Frank J. *From Five Fingers to Infinity: A Journey through the History of Mathematics*. Chicago: Open Court Publishing, 1994.

Taton, Rene, ed. *The Beginnings of Modern Science*. New York: Basic Books, 1958.

Trefil, James. "Puzzling Out Parallax." *Astronomy* 26, no. 9 (1998): 46–51.

Walker, Peter M. B., ed. *Chambers Dictionary of Science and Technology*. New York: Chambers Publishers, 1999.

Westfall, Richard. *The Life of Isaac Newton*. Cambridge: Cambridge University Press, 1993.

Whipple, Fred L. *The Mystery of Comets*. Washington, DC: Smithsonian Institution Press, 1985.

Wilson, Robert. *Astronomy through the Ages: The Story of the Human Attempt to Understand the Universe*. Princeton, NJ: Princeton University Press, 1997.

Wolf, A. *A History of Science, Technology and Philosophy in the 16th and 17th Centuries*. New York: Macmillan, 1968.

World Book Multimedia Encyclopedia. Chicago: World Book, 1998.

Internet Sources

About Water Levels, Tides and Currents. National Oceanographic and Atmospheric Association (NOAA), 2000. ⟨www.opsd.nos.noaa.gov/about2.html⟩

Britannica.com. 2000. Britanica.com, ⟨www.britannica.com⟩

The Galileo Project. Albert Van Helden, 1995. ⟨es.rice.edu/ES/humsoc/Galileo/⟩

Out of This World: The Golden Age of Celestial Atlas. Linda Hall Library, 2000. ⟨www.lhl.lib.mo.us/pubserv/hos/stars/welcome2.htm⟩

Treasure Trove of Scientific Biography. Eric Weisstein, 2000. ⟨www.treasure troves.com/bios/⟩

A Walk through Time. National Institute of Standards and Technology, 1997. ⟨physics.nist.gov/GenInt/Time/time.html⟩

SUBJECT INDEX

Subjects that have major entries in the text are indicated by **boldface type.** For publications, author names are indicated in parentheses.

NAME INDEX

About the Author

MICHAEL WINDELSPECHT is an Assistant Professor of Biology at Appalachian State University in Boone, NC.